The Sound of Feathers

Kathryn Gillespie

THE SOUND OF
FEATHERS

ATTENTIVE LIVING IN A WORLD BEYOND OURSELVES

DUKE UNIVERSITY PRESS
Durham and London

2 0 2 6

Project Editor: Liz Smith
Designed by A. Mattson Gallagher
Typeset in Garamond Premier Pro, Retail, and Forevs by
Westchester Publishing Services

Library of Congress Cataloging-in-Publication Data
Names: Gillespie, Kathryn (Kathryn A.) author
Title: The sound of feathers : attentive living in a world beyond ourselves /
Kathryn Gillespie.
Description: Durham : Duke University Press, 2026. | Includes
bibliographical references and index.
Identifiers: LCCN 2025020360 (print)
LCCN 2025020361 (ebook)
ISBN 9781478032861 paperback
ISBN 9781478029403 hardcover
ISBN 9781478061618 ebook
Subjects: LCSH: Human-animal relationships—Moral and ethical
aspects | Human-animal relationships—Social aspects | Animal welfare—
Moral and ethical aspects | Animals and civilization | Animals—
Effect of human beings on
Classification: LCC QL85 .G555 2025 (print) | LCC QL85 (ebook) |
DDC 591.5—dc23/eng/20250909
LC record available at https://lccn.loc.gov/2025020360
LC ebook record available at https://lccn.loc.gov/2025020361

Cover art: Adele Renault, *Gutter Paradise #11* (detail), 2019.
Oil on linen, 78 × 110 inches. Courtesy of the artist.

For Peter

Contents

Preface

Bringing this book to publication has been a painstakingly slow process, beginning in 2016 and continuing through the years of the COVID-19 pandemic. The book itself has changed during these years, not only because of catastrophic world events like the pandemic and the deepening climate crisis but also because I, myself, have changed. For me, these years have been characterized by an intensity of transformation and growth I hadn't experienced before in my life. I ended a twenty-year relationship and moved from my urban life in Seattle to a town of around a hundred year-round residents in a remote part of Washington State to follow my lifelong dream of living on the land.

During this time, I've built a home here from the ground up on fifteen acres that had a well but no other infrastructure, struggling through learning everything—How to frame in walls in a shipping-container-turned-art-and-workspace. How to make flooring that looks like hardwood out of plywood. How to build a staircase up a steep hill. How to build a fence to keep deer out of the garden. How to melt four inches of ice on the roof to install a chimney for a woodstove. How to build a chicken coop in zero-degree weather. I've learned which plants are noxious weeds that I'm legally required to remove, and I've started to learn which plants are native to the area and have medicinal qualities. I've learned to make salves and tinctures and put up food for winter. I've learned to buck hay. I've learned to "bed" a water line in a six-foot-deep trench, and I've learned how to dig my car out of snow drifts when I've driven off the road in icy conditions. I've learned to sit perfectly quiet and still to befriend the chipmunks here (I've just paused typing this because one is sitting on my thigh at my desk).

I'm somewhat accident-prone, and this place is rough for someone with that tendency. I've fallen off roofs and ladders. I've smashed fingers with rocks and hammers. I've nearly knocked myself out getting plywood out of the back of a truck. I've shocked myself on faulty wiring. I've been cut and gouged removing rusty barbed wire. I have scars from cuts and scrapes too numerous to remember how I got them. I've been sent flying on my face when I was rammed from behind by the neighbor's wandering horned sheep. I've dug too many holes in hardpack dirt and rocks to count, my back killing me, screaming my frustrations in a string of profanities across the valley.

I've struggled to maintain my sanity through winters of minus twenty-eight degrees Fahrenheit and feet of snow for six months of the year. My pipes have frozen and burst more than once. I've had hot water off and on (more off in that first year). I've gone for months at a time without a working shower and had the best date of my life because of its absolute perfection under the circumstances—coin showers at the gas station and burritos from the taco truck in the town forty minutes away.

It's not an easy place to live. In fact, it's probably difficult more times than not. But there's something electric and magical in making a new life—really making it, with my hands and muscles and sweat and grit and, yes, a lot of tears and rage. There's something beautiful, too, about life being dictated by the seasons—by planting and harvesting, preparing for winter, waiting eagerly for spring. There's conversation (and sometimes disagreements) about when it's safe to put your tomatoes outside in the garden or whether the first frost will come late enough for them to ripen. Everyone works on getting their firewood for winter and splitting and stacking it in late summer. There's hay-hauling season to ensure food for the animals in the winter. There's garlic harvesting in July and then planting in October. There's noxious weed pulling when the musk thistle and hoary alyssum are coming into bloom. There's a communal aspect to all of this, even when you don't talk to people about it and even when there's divisiveness within the community (and there is that, in spades)—everyone doing the same tasks on which life relies at the same time. There's also beauty in the simplicity of life lived in this way that lends itself to a slower, more attentive way of seeing and being in the world.

Here, I've also learned to love in a new way, a perhaps truer way, a way that's different from the kind of love we fall into at eighteen or twenty or thirty, the kind that navigates the wounds and experiences accumulated over decades that have shaped how we arrive at who we are today. James, my partner in this new experience of love, lives across the road, working on his own project of building a life here in the place where he grew up after a long time away. We coordinate our planting, share our harvest, and shoulder together the difficulties of living here. We help each other with two-person jobs and learn from each other, creating an intimate sense of community tied to the cadence of the mundane and extraordinary moments of everyday life.

In a lot of ways, I'm a different person now than the person I was when I first drafted this book eight years ago. But what hasn't changed—in fact, what has only intensified—is my interest in and dedication to practicing attentiveness to the world around me. Here, there is more wild nature than human infrastructure, and it means learning to respect, be in awe of, and sit in a near-constant state of unknowing about the world around me. It's a place of wonder in its newness, its strangeness, the hope of learning more, and the many things it's impossible to know. It makes me feel alive to try to cultivate an openness to being shown just how much I don't know, and to revel in the tiny secrets this place and the inhabitants here (human and nonhuman) have to share.

It was my dad who brought me here. He moved here during the pandemic to live with my uncle, tucked away safely in such an unpopulated area. Two years later, I followed him. He had helped me fall in love with this place when I'd come to visit and we drove around marveling at the landscape, working together to harvest garlic, raspberries, and strawberries, doting on Jinsky the tabby cat, and sitting in the backyard at night, looking up at the stars. So many stars in this dark, dark sky.

A year after I moved, he died unexpectedly, knocking my world off its axis. He didn't get to read this book—something he was waiting eagerly to do. But in my unfortunate perfectionism, I wanted it to be better than it was before he read it. Then, he fell and hit his head, causing an irreparable brain bleed, didn't regain consciousness, and died three days later. For those three days, my sister and I sat at his bedside talking to him, telling him jokes and stories and sharing what he meant to us, playing his favorite music, relaying the latest town gossip, and reading. He taught me how to write—with love and dedication, and with a brutally cutting criticism that would crush and inspire me in equal measure. Ours was a shared lifetime of exploring what writing is and can be. It was a shared project of choosing words carefully, of thinking about their meaning, what they're saying, and how to articulate thoughts that can feel impossible to put into words. The way his mind worked was wholly unique and rare, and he would often say, in an offhanded way, something so staggeringly insightful, so strange, so extraordinary it took my breath away. As he was dying, I read to him from this book, hoping that he could hear the words and know that this was the gift he had given me.

This past year without him has been one of unmooring, of living on unstable ground, stabilized in moments by the connection to the ground beneath my feet, the dirt under my fingernails, the lush green of the hills in spring and their burnt gold in late summer, and in the rhythm of life unfolding for the deer, elk, chipmunks, ground marmots, cows, sheep, coyotes, bobcats, and so many others. The sound of the wind in the aspen and the songs of birds in the garden. The static electricity and smell of rain in the air as a thunderstorm rolls in. The absolute silence of fresh snow. There's incredible stability in being on the earth and remembering that we're also of the earth.

I share all of this to give context for the life I'm living, the stories I tell in this book, where I've been, where I am now, and the unknown of the futures to come. Our lives are lived in stories, in encounters that shape who we are, how we see the world—what is, what was, and what could be. The invitation I open to you in this book is to look at your own life—who you are, where you are, and who you want to be—in new ways. I invite you to wonder. To look with a new attentiveness at what and who is around you, and to think about what could be.

Acknowledgments

Many thanks to Elizabeth Ault for her editorial guidance and belief in the book, to Benjamin Kossak for shepherding me through the process leading to the book's production, and to Duke University Press for publishing the book in its final form. The first draft of this book emerged as a result of the generosity and friendship of Lori Gruen, who provided me the time and space to do this work through the Wesleyan Animal Studies postdoc. There were many people who helped not only with their support but also with their expertise and knowledge that helped form the research for the book. "The Scent of the Spectral" was the most technically difficult chapter to write, requiring meticulous research on the ins and outs of biomedical research and its use of animals. I thank Rachel Bjork for sharing her immense wisdom and experience from years of familiarizing herself with this world and her dedication to shedding light on the experience of animals in labs, and Ryan Merkley from the Physicians Committee for Responsible Medicine for his technical support and information. I workshopped "Ghosts in the Garden" with the Animal Studies Working Group at the University of Washington, gathering brilliant feedback from María Elena García, Sarah Olson, Mei Brunson, Tony Lucero, Louisa Mackenzie, Glenda Pearson, Jessica Holmes, and Alyne Fortgang.

María Elena García has been a continuous support and inspiration across our fifteen-year friendship and colleagueship, reading and offering her feedback on the full manuscript and engaging in so many intellectually stimulating discussions about human-animal relationships that have pushed me to think more critically and carefully about these topics. To Rosemary-Claire Collard, a comrade and wonderful friend, for both reading the manuscript and giving me feedback, inspiration, and gentle nudges when I needed them. To Eric Haberman, whose support over the years helped to give me the time and space to write and think. I'm certain I wouldn't have been able to bring this book to publication without Yamini Narayanan's endless support—she not only gave such thoughtful feedback on multiple versions of the manuscript but also talked for hours and hours over years with me about the book, and has been the book's most enthusiastic, loving, and caring cheerleader at times when I had thoroughly given up.

To Lucy Gillespie, for her beautiful sistership as we navigate the world together and for her quantitative and demographic wizardry on this project. To Anne Franks for reading this so early on and giving her brilliant feedback, and for her unending support of this book, my writing, and my life at a deep and true level. To Rick Gillespie, who has helped me feel like I belong in my new home and whose care and intellectual companionship have become dear to me. To Laurel Rayburn, whose incredible mind, heart, and friendship feel like the essence of home. To the beagles Saoirse, Lucy, and Amelia for all they've taught me about the power of intense loving connection with members of other species and its capacity to heal deep wounds. To James Petty, my partner, my love—whose care and brilliance as a partner and as a reader of this manuscript gave me the last push (emotionally and intellectually) that I needed to transform the book into its final (and much better, as a result) state. He saw things I couldn't see about the book, inviting me into a new kind of attentiveness and way of knowing and thinking about my own life and work. And, finally, to Peter Gillespie, who taught me how to write, to think, to find startling new ways of understanding words and the world. His gifts of writing, thinking, and teaching stretch across a lifetime, living beyond his death and illuminating a path forward for my own continued life of writing.

Introduction

LISTENING FOR THE SOUND OF FEATHERS

When the wind catches in a particular direction and blows the marine air into Seattle, the morning smells to me like life itself. On early morning walks with the three beagles with whom I share a home, late in the chilly, wet Seattle winter, the world is vibrating with life. Bright green mosses cover walls and paths, and ferns unfurl new fronds. A misty rain moistens every surface, slowly creating drips from the few remaining giant cedars. Droplets form on every blade of grass, the damp soil and decomposing leaves mixing with the briny Puget Sound air—an intense fusion of earth and sea.

Snails creep across the sidewalk in the hours before dawn, hopefully tucking themselves in somewhere safe before the commuter foot traffic threatens to crush them. A flicker uses their beak to scratch for a breakfast of insects in a sidewalk crack. A pair of Steller's jays swoop, two brilliant blue flashes, screeching their unmistakable screech, past a lone crow sitting on the top of a telephone pole, peering down at life unfolding on the ground. A rabbit pauses in the tall grasses of an unmown lawn—alert, ears twitching ever so slightly. It's garbage pickup day, and a pair of racoons lumber away from an overturned trash can, garbage strewn across the street, and disappear into one of the few remaining overgrown lots, dense with ivy, underbrush, and trees. Are they just now heading to bed? A trio of ducks kept in someone's side yard with a scuzzy artificial pond wake up as I pass by, shaking their tail feathers at me and quacking their morning greeting. Somewhere down the block, I hear the familiar clucking of chickens as they awaken to begin their day in someone's backyard. The beagles I'm walking with are alert, listening, alive as their noses skim the ground, following a scent, their tails up and their paws wet from the damp grass of the lawns on which they trespass in these early morning hours.

There aren't many other humans out at this hour: a few early morning dog walkers who might stop for a quick canine hello, or who might hurry past, choking up on their dog's leash, pretending not to see Lucy the beagle prancing in eager anticipation of a canine encounter. A lone runner jogging methodically in the direction of the park. The twentysomething heading home wearing last night's clothes, not in a hurry. An elder out early, pruning a bush in their garden. But mostly, the world of the human city is still asleep.

I walk quietly down the sidewalk, listening carefully for the sound of feathers. Some days, one of the crows I know swoops down low, from behind, and flies past my ear—the whisper of feathers tracing a light wind across my cheek. They land on the power line above where I'm walking with the beagles. Together, we wait while Lucy, Saoirse, and Amelia sniff a spot of shared interest in the grass. We continue on our walk and make it halfway down another block when I again hear the rustle of wings in flight close overhead. Again, they alight on the power line. We walk like this, through the neighborhood. Swoop, rustle, whisper, wait.

Sometimes, we're out on our walks and this particular crow comes to get us. I'll be daydreaming, my gaze half focused on a plant I don't recognize in someone's garden many blocks from our house, when I feel the presence of someone above me. I look up and there they are, peering down at us, sometimes offering a little chirp in what I've come to understand as a *hello*, or maybe a more casual *hey*. They follow us home and wait while I go inside, unleash the beagles, and grab a handful of almonds or peanuts. I bring them out and toss them into a little green dish I keep on the front walk beneath our fig tree for sharing food, on occasion, with the crows.

This relationship started when, one summer, I noticed a pair of crows raising their fledgling on our block. As I would later learn, there is much about observing crows that requires a level of attentiveness that I'm not sure is possible in most humans. But noticing a baby crow learning to navigate the world does not require any level of refined attentiveness. They are incredibly loud. Their demanding, croaky cries are unmistakable. They wait on the ground, or perched on a wall, crying, until their parents appear with some food, at which point the cries intensify as they tilt their heads back, extend and flap their wings slightly, and wait for the food to be delivered into their bright pink mouths. Then they are suddenly silent, gobbling down whatever treat has been gathered for them. In that moment of eating, they look like a giant version of any other baby bird. But at any other time, to the undiscerning eye, they look like an adult crow.

By the time we see fledgling crows, they are nearly full grown in size, but they usually spend most of their time on the ground, which can seem like unusual crow behavior to the human onlooker. This is perhaps one reason why so many people mistakenly believe fledgling crows are injured and try to "rescue" them by capturing them and taking them home or to a wildlife rehabber. Late spring is what a friend of mine refers to as *crow-napping season*. In most cases they are not, in fact, in need of "rescue" or any help at all—they are simply learning to live in the world, and if you're attentive to the context, you're likely to notice that their parents are close by, watching them from a distance, feeding them, protecting them, and teaching them to find food and eat on their own.

It was this particular fledgling crow, whose parents watched them so carefully, who would grow up to come along on our dog walks. We spent that first year getting to know each other. I sat for hours watching them, wondering at their lives and lifeways. I read as many crow books as I could find. I tried to understand them. And I felt persistently—exhaustingly—inept. It took me months to even be able to wager a guess at who was who in this little group of three. I studied their faces the best I could. I looked at their feet and their wings, the shape of their beaks, their feathers, the way they behaved in their familial structure of three. There were things I noticed about their behavior, or I'd notice that one had gotten something white (bird shit, probably?) on the end of one wing, or that another one was molting faster than the others. But in the end, the only way I could reliably identify them was by the most rudimentary observational quality: their size. There was one parent who was the biggest, the other parent who was slightly smaller, and the fledgling who, although they grew similar in size to the smaller parent, was in that first year slenderer. But even these differences were only discernible relationally—when all three were there together and I could easily compare their sizes. If they showed up alone, most often I would be back at square one.

By contrast, crows have incredible facial recognition, remembering the face of a human person for years after an encounter with them. In a study at the University of Washington, conducted on the capacity for facial recognition in crows, researchers donned what they termed "dangerous" masks and then captured crows on campus.[1] These crows were then held captive in labs for a period of time, studied, and then released. Long after their release, crows in the campus community would harass researchers when they put on the masks. They remembered that these particular faces had been the ones to capture them, and they retained that negative association.

When I read about this, I wondered what it would be like to be trapped, held captive, studied, and released—the uncontainable panic and fear the birds would experience, both the ones being captured and those watching from the safety of a tree or wire. The capture of birds and other wild animals, including the practice of fitting them with bands and tracking devices, is highly routine in both academic research and conservation work. These practices are framed as being for a greater good—of counting or following the individuals of a species, of studying the bodies and lives of birds, of obtaining greater knowledge about other species.

In my experience trying to understand the crows, and in my research with other species, I've found that the pursuit of knowledge (my own and others') is powerful and often can result in an incursion into animals' lives and homes. It's difficult to read a book about crows or any other species and realize that some, if not most, of that knowledge was obtained through some form of disruption or harm (temporary or permanent) to individuals. It's difficult especially because of the joy and wonder that learning about other animals can evoke. Reading about crows, I was at once filled with gratitude for this knowledge and a pang of pain in thinking about how it was gotten.

As that first summer passed, my relationship with these three crows evolved. Some mornings they would show up, demanding to be fed. My bedroom window was just above the front walk, and they discovered that if they came around at six o'clock in the morning and rattled the plate on the cement walk, it made such a loud racket that I would eventually get up to feed them. While one waited in the fig tree, another rattled the plate, and the third would sit on the power line that stretches from the street to the front of our house. At this level, they could peer directly into the bedroom window. As soon as I sat up in bed, the crow surveilling the bedroom would perk up, stare straight at me, and caw loudly, insistently. The crow on the ground would bang the plate again. I would get up to feed them. I'd been well trained.

Being in relationship with these crows has made me attuned to other crows. It's odd to think of my intentionally cultivated relations with these crows suddenly making me aware of crows everywhere I went. Hadn't I, in fact, been in relationship with these and other crows for years? Hadn't I, in fact, been living in relation with all the other human and nonhuman lives around me since I was born? But I didn't register crow calls or notice the magical sound of wind through their feathers as crows fly past until I rendered myself attentive to their lives, their bodies, and their presence.

That first summer, as I brought my attention to the crows, during molting season, I noticed that the neighborhood was covered in crow feathers. As it turns out, it's like this every year, but in previous years I hadn't noticed. One afternoon, I was walking home, and as I came up to our house, I found a long, shiny-black crow feather on the sidewalk. I picked it up and twirled it around between my thumb and forefinger, admiring it. One of the crows chirped from the electrical wire above. I looked up. They peered down, watching. I gave them a nod and took the feather into the house and put it on a shelf filled with treasures like fallen feathers, rocks, seashells, and pine cones.

The next day, when I delivered the crows a treat of sunflower seeds, I moved back a few steps from the dish and watched. The largest of the three crows swooped down and landed with one claw on a low branch of the fig tree in the front yard. In the other claw, they clutched a perfect crow feather. The crow looked at me. Looked down at the plate. Looked back at me. And then slowly extended the claw holding the feather and released their grasp, dropping the feather right next to the plate. The crow looked at me again and cawed. I stepped forward carefully, and they hopped up higher in the fig tree as I moved to retrieve the feather. I picked it up and looked closely at it, appreciating its beauty. I whispered *thank you*.

I walked back to the house, wondering with awe at the gift I'd been given, and I heard the familiar rustle of feathers as the crow dropped down to the ground to discover what snacks I had left for them. *Remarkable,* I thought. Noticing I appreciated the beauty of the feather I found the day before, here I was delivered another even more stunning feather by one of the crows themselves. Was this one of their own feathers, I wondered? Was it a feather from a stranger that they had found on the ground and brought here? These kinds of things are probably not knowable, but I like to spend time wondering.

* * *

The following spring, the crows disappeared to build their nest, lay their eggs, and tend to their young. Often, a pair of crows will be joined by a third crow in caring for the nest and parenting, and usually a young crow will stay with their parents for a year or two before finding a mate themselves. I wondered if the young crow born the year before had joined their parents in co-parenting whoever was incubating and hatching in their nest.

After what felt to me like a long wait, joyously, they emerged from the nest with three healthy fledgling crows. The family of three had become a family of six. Our garden suddenly felt like a proper gathering place for crows—all six would descend into the fig tree or land on the fence, hopping deftly down to the ground. The three young ones were awkward and timid, watching the adults carefully and taking cues from them about when it would be safe to enter the yard and when to spring into flight and move to safety.

The sound of feathers came in a wave that year the crows descended, a flurry of black wings. There were times when I worried about how conspicuous they were visiting our yard. When there were three, they had been noticeable if you were a neighbor paying attention, but when their flock grew to six, they were impossible to miss. On the neighborhood social media app Nextdoor there had been a series of posts around this same time about the fact that crows, and other birds as well, were digging up people's lawns to get at the tasty grubs beneath the grass. I had seen them doing it all over the neighborhood. The lawns were shredded, little tufts of dirt and grass yanked up and tossed aside.

Overwhelmingly, the consensus was that crows were destructive, loud, dirty, and aggressive, and many people claimed they had no business being in the neighborhood. "They're just up to no good," one commenter posted. Others suggested all kinds of possible solutions to driving the crows from the neighborhood and addressing the grub-digging problem in particular. The most common seemed to be to "get to the root of the problem" and poison the grubs, followed by some suggestions to poison the crows themselves.

Nextdoor is meant to be a social media platform dedicated to bringing neighbors together around shared conversations and issues, but what I've observed over the years reading posts and comments is that it is often a space for grievances in various forms to be expressed and validated. Disgruntled complaints are targeted at all kinds of things: at airplane noise under the flight path, at people who use leaf blowers, at car drivers by bicyclists and at bicyclists by car drivers, at dog walkers who don't pick up their dog's poop and at dog walkers who do pick up their dog's poop but toss it in other people's garbage bins. I've noticed, especially as it relates to topics like the poisoning of crows or the eradication of coyotes or raccoons (or any other number of undesirable urban species), that the online space allows for an echo chamber of validation for increasing intolerance of other species. And this intolerance is not limited to the digital sphere but travels out into the world to harm the animals themselves—the crows, the grubs, the rats, and many others.

Both online and offline, many people seem to view crows as a nuisance for their propensity to get into garbage bins and make a mess, or because they're loud, or for whatever other myriad reasons might attract disdain. But in this case, the fact that they were now destroying people's lawns—the carefully manicured pinnacle of the middle-class landscaping aesthetic—moved them into a category marked for eradication. I worried over this little family of six—that they were attracting attention here in the yard. I worried that the crows would be poisoned while they were out foraging for the day, and I was relieved every time I saw them show up, together, as an intact family.

Crows flourish in areas with high densities of humans. Unlike other species who may be driven out of cities by human incursion into their homes, crows thrive in human presence. They are scavengers and find plenty of food discarded by humans, and they make their homes in places shared with humans and with other urban animals. But it is precisely this flourishing that marks them as maligned—as a problem to be addressed. As I observed this family of crows over months and years, I wondered: What might it look like to hold in our hearts and minds the histories and lived experiences of those with whom we live closely but who are targets of eradication and expulsion? Or even those we may just ignore in their mundane ever-presence?

It was sometime in the fall of the three new fledglings' first year that I noticed one of them had light-colored crusty spots around their beak and eyes, and that their feet had some growths on the bottom and on the toes. I watched the crow try to land on the ground and immediately hop up as if the impact was painful. I learned from research that it was likely avian pox and that often it could resolve on its own. So, I watched the crow carefully and observed the rest of the family to see if the disease would spread to anyone else, but all the others seemed to be fine. To my untrained eye, it finally looked like the lesions and scabby spots were clearing up, but a week later the crow disappeared. The family of six turned into a family of five. I assumed the crow had died, and I wondered where they had died and if the family had been there with them. I had read that crows mourn the death of their kin, and I contemplated this family's grief and their rituals around death. I wondered about the mysteries of their experience of loss and mourned for what they might be feeling.

After this disappearance, I watched with a little sadness each evening as they joined with flocks of other crows and flew to their roosts. During their nonbreeding months, crows gather each night to roost in the thousands in a chosen meeting place. These crows and I lived in the south end of Seattle. I had heard from a friend who lives in the north end of town that the crows there fly to a shared roost near her home. The crows in our neighborhood flew south instead of north. I researched online and learned that there is another roost of thousands of crows in an office park next to an IKEA south of Seattle. As dusk fell, the sky was dotted black with crows all flying perfectly in the same direction, cawing to others as if to say: *It's time. Let's go.* I wanted to follow them, to see their roost—to witness the collective gathering, the cacophony of chatter as they came together. The impulse was strong to see what this was like, to try—likely with futility—to understand this family's connection with this much greater whole. But I also wanted to respect their space and this time. They had chosen to spend part of their day with me, here in our shared home, and now they were choosing to end their day in the company of other crows.

There is so much to notice on the everyday walks I take. So many details of life and death unfolding in block after block of urban sidewalk. This noticing requires a level of attentiveness to tiny details—moss in a crack on a stone wall, a beetle overturned and dying, dew on a spider web, the high-pitched chirp of a hummingbird, the smell of decomposing leaves, the sound of feathers. This attentiveness necessitates slowing down—not only in pace but also in thought, clearing space in the mind for wonder. To slow down, notice, and wonder often feels like a reminder of how vibrant the world is, and of how rushing through each moment, hour, and day can cause us to miss many of the most magnificent mundane details of the world. There are moments on these walks when I'm overwhelmed by this beauty and liveliness, by the sheer joy of being present for what is. And still, I'm missing so much. I'm missing so much that I don't even know what or how much I'm missing. It's a constant struggle to pay attention, to try to anticipate what I might not notice.

What does it mean to *pay attention*? In her book *How to Do Nothing*, Jenny Odell writes, "If we think about what it means to 'concentrate' or 'pay attention' at an individual level, it implies alignment: different parts of the mind and even the body acting in concert and oriented toward the same thing. To pay attention to one thing is to resist paying attention to other things: it means constantly denying and thwarting provocations outside the sphere of one's attention."[2]

To pay attention is a sensory experience—it is not just a matter of looking at something closely. It involves using the senses available to us to fully experience the subject of our focus. With sight, we might take in the lush greens of the moss and cedars and ferns. With hearing, we might listen for the *caw caw caw* of a crow. With a sense of touch, we might kneel to feel the cool, damp blanket of moss and experience the moisture from the ground soaking through the knee of our jeans. With a sense of taste, we might tilt our heads back, open our mouths, and wait for a raindrop to fall. These forms of attention combine in whatever forms of sensory experience shape our lifeworlds, crafting a fuller practice of noticing the cold, misty Pacific Northwest morning.

There's also what's behind the thing we're noticing. Alexandra Horowitz, in her book *On Looking* about the act of paying attention through everyday walks in her neighborhood, explains that "part of seeing what is on an ordinary block is seeing that everything visible has a history. It arrived at the spot where you found it at some time, was crafted or whittled or forged at some time, filled a certain role or existed for a particular function. It was touched by someone (or no one), and touches someone (or no one) now."[3] I'm interested in these histories—the material histories of objects and subjects, and the larger structures in which these acts of noticing, daily life, beauty, and destruction are formed. Attending to something we encounter every day—noticing the thing that is the opposite of noticeable—can reveal rich and complex histories that shape the present and which connect that thing to different times and different places.

Odell's caution that noticing one thing means ignoring or forgetting or missing other things complicates what it means to pay attention. I bend over to peer at an earthworm struggling to slither across the pavement. I consider whether to pick up the worm and move them to safety in the direction they were heading, or to let them be. While I contemplate the worm, I might not notice a squirrel leaping from one branch to another in the tree above my head. I might not notice a car running a stop sign at the corner. I might not notice Amelia devouring half a bagel dropped on the parking strip before the other beagles can get it. What other hundred or thousand things am I missing in my focus in that moment on this singular earthworm?

The noticing itself can also obscure a way of understanding this multispecies world that offers a fuller story of the broader picture of life and death today. It may be easy to imagine, for instance, with focused attention trained on the lush green of the Pacific Northwest city vibrating with multispecies life, that the world as we know it is not dying. That the climate crisis is less urgent than it is or that it's not something to worry about right now, at this moment. That this single, classically Pacific Northwest damp, drizzly winter morning is not, now, an anomaly, sandwiched in between months on end of unusual heat, drought, and smoke in summer and torrential rains that bring winter flooding. Noticing, then—paying attention—can be dangerous. It can distract our attention away from the things that may need it most—the things to which we desperately, perhaps for our sanity, do not want to attend.

There's an easiness to appreciating beauty. And there's even an easiness to paying attention to that which does not necessarily elicit feelings of joy or pleasure. An easiness to noticing things that may prompt disgust: the pungent smell of a rat's flattened corpse rotting in the gutter along the street. A hypodermic needle or a used condom tossed carelessly in an alley. A crow pecking at a soiled tampon dragged from someone's trash. But there is no easiness to attending with clarity and persistence to the things that cause us pain, grief, shame, regret, or fear, and there is much to observe in the world that provokes those responses. To pay attention can be an emotional experience—part agony and part euphoria in equal measure.

You look at a blade of grass, soaked with the morning mist, and you might think, "Look! Some grass, wet from the rain!" But what is the fuller story of that blade of grass? We might think about its history, how that grass came to be here, as part of this lawn, normalized as the kind of greenery that is desirable to urban and suburban land occupation. We might think about the chemicals that have been deployed to keep it so perfectly green, so immaculately free of weeds. We might consider whose labor keeps this grass neatly trimmed, a uniform mat stretching out in front of the house. Or we might reflect on whose presence—human, animal, plant—the planting of this lawn necessitated the displacement of. This process of attentiveness and the seeking out of knowledge to understand it might involve moments of discovery and surprise, and it might be deeply unsettling, destabilizing the things we think we know, the things we want to believe about the way we live our lives and the way we see the world. This unsettling—the feeling of being unsettled, the unsettling of norms—is at the heart of this book.

When I was little, a sense of wonder came easily and powerfully. As children, our lives are full of experiencing things for the first time, of learning what and who things are, how they're in relation with other things and beings, where things come from, and how it all fits together to construct social and environmental life. I remember as a kid lying flat on my stomach on the warm summer sidewalk watching a line of ants carrying crumbs off to their home, amazed by their strength, their focus, and the way they worked in concert with one another. I remember watching the seemingly endless loop of feeding in a nest of robins—the parents leaving and bringing food back to the nest, the baby birds crying desperately, and then the sound of frantic gobbling when it was delivered to their open mouths. I remember watching seagulls at the beach in North Carolina swooping in to grab a bag of chips when a beachgoer's attention was turned somewhere else and a murder of crows ransacking dumpsters that radiated the nauseating smell of rotten fish. I remember standing on the back porch in the steamy heat of Pittsburgh summers watching the fireflies light up the backyard in irregular bursts of glowing light, and the sound of cicadas singing their chorus, leaving behind their exoskeletons on trees to be carefully plucked from the bark and inspected in the light of day. This curiosity was rooted in a struggle to understand—to observe, to ask questions, to practice greater attentiveness, to fail to understand, and to observe more closely.

I remember observing living beings and letting them be, and I remember touching them, capturing them, inadvertently harming them, thinking I could care for them in an artificial environment—the lightning bugs caught and confined to a jar. The butterflies I held in my cupped hands who died as I watched, having been handled too roughly. There were what we called potato bugs (also known as pill bugs or roly-polies) I plucked carefully from the sidewalk and collected in my hand so that when I opened it, they all unrolled from their scared little balls and swarmed over my palm. There were dandelion flowers torn from the grass to make flower crowns. There were crawfish or salamanders fished from a stream and played with until I got bored. And there were the tadpoles sloshing around in a jar as I checked every few minutes to see if they'd turned into frogs. Without direction otherwise, the urge to touch, to hold, or to investigate animals in captivity is an automatic impulse, without knowledge of what harm it can do.

There were also times when the harm was not inadvertent or unintentional, framed as fun, even as it was clear that the animals were suffering, as in the case of the live blue crabs my dad would buy from the fish store on vacation. We would hold crab races on the kitchen floor, watching as the crabs crawled from one side of the room to the other before we boiled them alive in a large pot, holding the lid down tightly to stop their efforts to crawl out. I remember, too, perching on the edge of a dock with our fishing poles, stabbing the fishing hook through wriggling worms again and again to make sure they were secure bait. I recall the excitement as we hauled a fish out of the water as they thrashed in the air, and the frenzy as we struggled to tear the hook out of their mouths before tossing them into a cooler. The ambivalence I felt in these moments was a mix of flurried excitement and distress at these animals' obvious suffering.

I was a sensitive kid, heartbroken over the tragedies of nature or the overlooked pain of those in this multispecies world around me. I was also—as most kids seem to be—insatiably curious. I was in a constant state of wonder, of wide-eyed joy and surprise at learning something new—a bottomless well of questioning and craving satisfying answers. I (and I think a lot of others) experienced childhood as having endless questions that bubble up and need answering to understand the world. Things are magical, observed for the first time. As we age, we think with increasing confidence that we understand the world around us, that we've learned a lot about what things are and how they work, or maybe it's just the exhaustion and responsibility of adult life that co-opts the time we might otherwise spend in this wonderous state. Whatever it is, in the process, we lose some of the joy of those first discoveries of observing closely, seeing the enchantment of the places we live. Our lives become routine. The sense of wonder slips away, little by little. And we can be alienated from the nuanced details of the world around us.

How, as we get further from childhood, do we reclaim that buried sense of wonder? How do we cultivate a renewed attentiveness, looking at the world with fresh minds and hearts to be electrified by the magic of what's around us?

As I've gotten older and near middle age, I've noticed that I and other adults have come to accept different kinds of harm to other animals and the environment as unavoidable or intractable problems and just the way things are. We come to know our place in the world—what's familiar, what we'll tolerate, what we'll ignore, what we'll devote our love and attention to, how our living involves accepted kinds of harm. The harms we delivered on other species as kids, without an awareness of what we were doing, evolve into other routine kinds of harm—some of which I explore with you in the pages of this book—and I invite you to consider the kinds of observations that I share that have evolved over my life and brought me to the particular questions I have now, questions that have both come out of and led to a certain attentiveness.

How can attentiveness to our everyday encounters with other animals help us to understand the relationships of harm and care in which we're embedded whether we've noticed or acknowledged them or not? How can we look askew at mundane and taken-for-granted relationships with animals—running over them on roads, eating them, testing on them in labs, exterminating them—in order to understand the fundamental nature and effects of these relationships in a new light?

Practicing the attentiveness needed to begin to answer these questions is a first step. After paying closer attention to the smallest things or beings in the world around me—making attentiveness a *practice*—my days now are filled with "*I wonders . . . ,*" an at-times relentless flood of curiosity and questions.

I was finishing writing the draft of this book as the third summer of listening for the sound of the crows' feathers came to a close and autumn began. Earlier that year, the family had disappeared as they do in the spring, stopping by on occasion, individually or in pairs, to see if there were any snacks to be had in the garden or to soak something they flew in from somewhere else (like a crust of bread or the foot of a sparrow) in the birdbath. But I still had trouble recognizing who was who when they weren't there together as a whole group. They were tending their nest, in some configuration of co-parenting, as had happened the year before.

I was left to wonder where their nest was, how they were faring, how many bluish-green eggs they had laid, how many would hatch, and how many fledgling crows would make it to adulthood. There was mystery and some magic, too, in their disappearance, and in the feelings of wonder, the not knowing, the quiet hope for their well-being and safety, the eager anticipation of their reemergence as a newly formed family.

I continued to leave treats in their bowl, and individuals of the family continued to stop by, but our relationship felt more distant to me. They were focused on other, more important things. As spring turned to summer,

I started to hear the unmistakable scratchy calls of baby crows around the neighborhood on my walks. I watched as other parents raised their young. I waited, eagerly, for the crows on our block to reappear as they had the summer before with their fledglings. But they didn't come. I waited. The summer passed, week by week, and they didn't reappear. I wondered what had happened. Had none of the babies survived that year? Had they decided our block was no longer the place to raise their young? Had something happened to the parents?

As the crows became less frequent visitors, other birds became more at home in our garden. The Steller's jays and the scrub jays—each their own hues of brilliant blue—swoop down into the yard to retrieve a peanut or an almond. They travel in troops—four Steller's jays or a trio of scrub jays suddenly appear in the tree all at once. The Steller's jays are more brazen, flying toward me the moment I step out the door, landing on a tree branch within reach, screeching their distinctive screech in my face. The tiny pine siskins and purple finches and wrens and chickadees come by and forage for bits of food on the ground and in the bushes. A single pigeon sometimes stops by, and I wonder about their flock. Squirrels hop confidently along the top of the fence, nibbling on the remaining apples on the apple tree and burying food throughout the garden. A rabbit squeezes under the fence, looking for some greens to eat. Someone comes by at night—a raccoon probably—and dirties the clean water in the bird bath.

Within this vibrant multispecies community, as autumn is in full swing and we move into winter, only a single pair from the family of crows remains. I wonder who they are. Are they the original pair of parents? Did the young crow from our first summer together find a mate and decide to settle here in this territory? Were these the surviving two of the three fledglings from our second summer together? Or were these, perhaps, someone else entirely?

Whoever they are, I am glad for their presence. But I also feel a sense of loss. I felt an intimacy with this evolving family of crows—that we were sharing food and feathers, yes, but also that we were sharing something more. Companionship on our walks. An exchange of *hellos* when I saw them foraging for grubs in the grass across the street. Long hours spent in the front garden watching each other, curiously trying to understand one another. I felt even a sense of intimacy in knowing somehow that they understood me so much better than I would ever be able to understand them. It was the feeling of being surprised to learn how well someone else knows you, how carefully they've paid attention to who you are, to what matters and has meaning to you. There is a great intimacy in that—in being known, in the attempt to be understood.

And yet, this intimacy with the crows came not from a deep knowledge of each other but from the consistent effort to try to know each other on our own terms—without my imposing on their lives, trapping them, banding them, keeping them captive in a lab, invading their territory, studying them, quantifying their behavior with meticulous scientific methods, as humans often do when trying to understand another species. What this family of crows and I crafted together was profoundly imperfect and unsystematic knowledge. It is overwhelmingly partial and entirely incomplete. They allowed me a glimpse of their world, and although I felt and still feel an intense desire to know more—to know everything about them—I think part of cultivating attentiveness is honoring the things we cannot know, respecting the secret lives of others all their own.

In this effort, how do we let attentiveness guide the work of gentler, more radically caring world-making? How do we come to know other animals differently, not through relationships of harm and violence but through tender acts of appreciation and care for what they might need to flourish—within, in spite of, and outside of interactions with humans?

Can we appreciate other species and accept that we may never fully understand them—and that our inability to fully understand them should not be justification for exploiting and harming them? Can we orient ourselves in and through the opposite: That we err on the side of respect for the radical alterity of their being, humbly recognizing our own limitations and our own lack of understanding? Can we sit, carefully, in a state of unknowing?

And ultimately, can we accept that even when someone else is unknowable—even when we can't even scratch the surface of what we think we can know—they are owed our full support in their flourishing?

These questions are grounded in the mundane moments of encounter we have with the world around us. The beginnings of answers to them, I think, can be illuminated by the tiniest, most everyday acts of attentiveness, of noticing, of working little bit by little bit on repair.

This is the project I've undertaken in *The Sound of Feathers*. The chapters that follow evolve across the fleeting encounters, the closest-in relationships, the entanglements of intentional harm, and the possibilities for greater attentiveness, care, and respect. Some of the animals I write about are ones I have encountered only in passing, but they have made their mark and shown me something important about the structures of power and violence the human world upholds. Other animals in the book are those I have lived with and been in complex relationships of care with for many years. And then there are those animals with whom I live in close proximity, not in intimate relationships of care but in ones of eradication and harm.

These encounters, although often involving painful stories about violence and the effects of power, raising fraught ethical questions, also offer storied windows, I hope, into new forms of care, flourishing, and multispecies kinship.

Careful attentiveness can enable this kind of storytelling. I hope this book offers a starting point for bringing attentiveness to multispecies relationships and thinking critically and carefully about the ways that we ourselves are engaged in relationships of harm and care. Perhaps reflections on our own implicatedness and our own critical attention to the ways we inhabit the world might help others to think differently about theirs. Perhaps, together, a tapestry of accounts of attending more carefully might help to transform our shared futures in ways we can't yet imagine.

I've done a lot of research for this book, devouring written accounts (both academic and nonacademic) of history, animal behavior and emotion, political economy—all kinds of topics that took me down rabbit holes of excavating written forms of knowledge, as well as fieldwork in spaces of animal life and death. This helps us understand the complexity of what we're seeing and where these relationships are rooted, historically and in contemporary society. Meticulous research is important and illuminates whole new depths of knowledge—it grounds our wonder in a set of existing knowledge that can guide our thinking and actions. But the truer and more fundamental element here is that act of paying attention, of developing greater sensitivity and curiosity about what's right around us. It's the spark of wonder that ignites our ability to begin to understand more deeply and narrate our connections with and our disconnections from others in caring, violent, or indifferent ways.

This is a storytelling practice and many of us know how to observe the world around us, to think about what it means, and to weave beautiful stories. As multispecies beings, living together in all this messiness, we are all the authors of these stories. We live in stories. Our lives are stories that unfold over time. All of us—human and nonhuman—know beauty and pain, love and sorrow, and rage and joy. We experience the toll that living in the world today takes on our bodies and minds and the way we are differentially affected by structures of power. We know what it feels like to be connected to others and to experience the loss of others. This is the stuff of what it means to pay attention. This is the substance of the stories we can tell, the questions we can ask, the more vibrant futures we can imagine.

1

Knowing Our Haunted Homes

> We howl in the dark for the loss that surrounds us now, and
> for all that is coming. Our howling starts from within, from
> empathy, grief, and much more, and it reverberates beyond
> us. At the same time, other howls reach us and penetrate
> us, amplifying not only our voices but our meaning.
> —Deborah Bird Rose, "In the Shadow of So Much Death"

It was one of those classically dreary Pacific Northwest mornings—foggy
and cold with a light mist that has a way of soaking through your clothes
almost without your noticing. We bundled up in rain gear and headed
north of Seattle to board a boat that would take us to see one of Puget
Sound's most iconic species—orcas. It was a lifelong dream of my mom's to
go whale watching, so while she and her partner were visiting from Pittsburgh for Thanksgiving, we set out to fulfill that experience on her bucket
list. We arrived early and found our seats on the boat. A naturalist was
our guide, and as we set sail, he began to tell us about the orcas and other
marine mammals in the area. He was an engaging speaker with a deep
well of knowledge and met our seemingly endless questions about the
details of orcas' lives with patience and detail. Most of the crowd seemed
focused on getting out to where we would see the orcas, and we took that

opportunity to raise our hands every few minutes with questions, excited to learn more.

The anticipation and eagerness of the audience were palpable, creating a frenetic energy on the boat as we neared the first pod of orcas we would see that day. Whale-watching companies in the region radio to each other the locations of orca and other marine mammal sightings so that every boat is successful in providing a viewing for tourists. We followed the lead from another boat and soon arrived at a pod of orcas. There were six boats gathered in the area. The crowds on each vessel clambered to the side of the boat that would allow the best viewing, pressing against each other to catch a glimpse of the pod. Loudspeakers echoed across the water as the guides on different boats described what we were seeing. People on one boat were throwing fish overboard to feed the orcas, encouraging them to be more active and come to the surface more frequently. Our guide spoke disparagingly of this practice, saying that it interfered too much with the orcas' lives and was something they felt strongly about avoiding.

There were three adults and a baby swimming in circles at the center of the collection of boats. Their black and white bodies flashed above the water, disappeared, and then reappeared. The jostling crowds on the boat snapped photographs of the orcas—proof, it felt like, that they had seen them at all. The naturalist, knowing that in the age of smartphones many people aim to catch nearly everything in our lives on camera, said, "Here you go, folks! Here's your photo op!" Some leaned over the edge of the boat, struggling to get a clear picture of the orcas as they emerged from the water only momentarily. Others turned away from the water, switching their cameras into selfie mode to try to capture their own presence alongside the orcas.

Years later, in trying to recall more detail about the actual orcas we saw, I could only come up with this most rudimentary description. I asked my mom what she remembered, and her recollection of the orcas themselves, too, was limited—she recalled that they were bigger than she'd thought they would be. She explained to me that it felt similar to seeing animals in a zoo. She had the sense that we were encountering them as objects of entertainment and performance, not as wild animals living their wild lives.

We watched for a short time before the orcas began to move away. Several of the boats followed them, but the captain of our boat said we were going to leave them be and seek out other animals for the rest of the tour. We saw dolphins who swam in front of the boat and seals and sea lions resting on logs that floated in the sun. The energy of the audience flagged after the orca sighting, and there was less enthusiasm while viewing these other species.

That afternoon, we walked off the boat with a feeling of disappointment and unease. We had been inspired and fascinated by the naturalist's presentation and his sharing of knowledge about orcas and felt a momentary thrill when we saw them. But viewing the orcas themselves also felt off, like maybe we weren't supposed to be seeing them. It wasn't the magic of spontaneously seeing an animal in their world, on their own terms, watching them pass in front of you as you marvel at the gift you've just been unexpectedly given. These kinds of moments, even if they're fleeting, engender a slowing down—an almost electric attentiveness to the animal you're observing, knowing that in an instant, they will be gone. The frenzy of the boats surrounding the orcas—the noise, the fish being thrown into the water, the loudspeakers and crowds talking loudly, the frantic snapping of photos—didn't allow for slowing down in quiet appreciation.

My mom talked about having seen whales at a distance from the shore of the Oregon coast years before. She described how, although she didn't get to see them as closely or clearly as on this boat where the point was an "up closeness," there was something different—maybe *purer*—about seeing them out there living their whale lives apart from humans.

The whale-watching experience was a curated search and viewing, a crowd of boats following and surrounding the pod, orchestrated with great effort—and at a profit to the whale-watching companies. All the companies I looked at guaranteed that you'd see orcas, whales, and/or other marine mammals while on the tour, and I wondered what the effect of this guarantee might have on the animals themselves.

We left the tour with much more knowledge about orcas than we had previously, but it was from the educational presentation on board and the personal experiences of the naturalist's own life, not from seeing the orcas themselves. What we learned from seeing the orcas was more about humans—the way our hunger to see and know other animals presses us into an incursion into animals' habitats, their movements, and the relationships they have. This desire *to know* comes, I think, from a longing for connection to other species and for an understanding of ways of relating that are both different from and similar to our own.

Often, though, we just don't know how to relate to animals in ways that support their flourishing. We sometimes encounter them as consumers—as paying tourists on whale-watching boats or as patrons at zoos and marine parks or in any number of other ways. I thought about this consumption as I watched the audience—seeing and hearing a sense of disappointment from other people on board and a feeling of being unfulfilled. I felt a hollowness coming out of the experience, and others, too, seemed to have had this same response.

One of the themes I explore in this book is what it means for other animals and for ourselves that our lives are defined by our consumerism, a relationship to the world around us that will feel familiar and which limits the depth of the kinds of relationships that are possible. Practicing attentiveness can help us intuit and understand the animals and environments around us, how they live, and what they need. But we also need a fuller and deeper understanding of *why* these relationships are the way they often are. And this requires uncovering the social, political, and economic structures that define and constrain our capacity to relate to others and cultivate more caring relationships.

To begin, what are these structures exactly, and what do they do? How do they shape the everyday multispecies worlds around us? Whose lives today are haunted by these structural logics, and how might we reckon with them in ways that lead to their undoing?

Two distinct yet intertwined systems form the foundation for how life and death for humans and other species unfold today—a capitalist economy and a colonial way of relating to the world. It is worth explaining briefly these fundamental structures, their histories, and their essential elements to offer context for the journey of discovery that unfolds in this book. Often, capitalism and colonialism are understood in primarily human terms in relation to inequality, struggle, and prosperity in historical and contemporary human society. But looking beyond the human, they also shape *all of life* in profound ways—both our human worlds and the worlds beyond ourselves. Many of the relationships we have with other animals are characterized by violence, but this violence is so insidious it doesn't seem like violence at all. The systems that normalize these forms of violence define the very essence of the kinds of relationships we do and can have with other humans, animals, and ecologies, and looking closely at them—really trying to understand *why* these relationships are the way they are—illuminates the oftentimes invisible power and shape of their impacts.

Capitalism organizes life around making the earth and its inhabitants into resources for production and objects of consumption, and at its core is the accumulation of capital based on these logics. The *commodity*—things, and living beings made into things—is at the heart of capital exchange and accumulation, generating value in its circulation.[1] Lives are commodified—in the making of orcas into sites of entertainment and profit, or in a more wholesale way, for instance, in the case of farmed animals who are bought and sold, and whose reproductive labor is commodified to produce new commodities to be bought and sold (as in the case of milk, eggs, and new generations of animals themselves). Death is commodified when their flesh is rendered into meat, and what remains after that is rendered into new commodities, like soap, pharmaceuticals, fertilizer, and a dizzying array of other products. Human labor is commodified. As humans, we sell our labor so that we can buy the most basic stuff of life—housing, food, clothing, medical care—all necessities that, because they are made into commodities, are unevenly accessible based in part on our geographic location and the wealth we can each accumulate.

The inherent drive of capitalism is perpetual growth, and for this to occur, production and consumption must continue and accelerate. This growth deepens inequality and intensifies what and how we consume. It's difficult to resist or exist outside of this cyclical growth. We relate to the world as consumers, and this dynamic alienates us from the deeper, truer elements of what it means to be alive and in relationships with others. I felt this in the hollowness of the orca viewing—in having sought out, consented to, and paid for what was an inherently capitalist experience of relating to the orcas' lives and world. As we rode on the boat through Puget Sound, traces of a capitalist economy were everywhere around us, not just in the whale-watching boats but in the massive barges carrying goods in shipping containers to and from East Asia, in the polluted waters of the sound, and in the distant skyline of the metropolis of Seattle. It was impossible not to see this defining structure around us if we were attuned to its presence and impact.

In trying to understand the nature of these relationships, capitalism doesn't work on its own to structure our social, political, and economic worlds. Colonialism is a defining system that shapes life and death globally, organizing the world along particular fault lines of power. In its essential definition, colonialism is the process of domination of other places, people, animals, and ecologies through the seizure of land, the dispossession of racialized people and animals from their traditional lifeways and the places they live, and the ongoing occupation by the colonial power. It's a process tied closely to imperialism—the conquest and domination of another place and people based on ideas of racial superiority, which doesn't necessarily involve direct rule but is characterized by economic control, military force, and/or political interference.

Settler colonialism—the kind of colonialism that characterizes societies like the United States, from where I'm writing—is a process through which European populations invaded and *permanently occupied* land stewarded by Indigenous communities and native animal species, transforming ecologies, murdering and displacing its residents, and growing the numbers of settlers to continue and intensify their presence and power. The evidence of the capitalist economy in and around Puget Sound reveals, if we attend to it, the arrival of Europeans in the area in the 1700s and 1800s; the occupation, displacement, and annihilation of what was here before; and the kinds of lifeways and forms of existence that were replaced by colonial ways of relating to this place.

Concepts of *private property* and *ownership* are central to this occupation, and the world has been ordered into what and who could be *owned* and who could do the *owning*. Owning and being property reflects a colonial ordering of life that differs starkly from Indigenous ways of relating to the multispecies world.[2] Colonialism hierarchizes life according to an *anthropocentric* worldview—the belief that humans are the central and most important figures in existence.[3] This can also be understood as a human exceptionalist or human supremacist orientation to other forms of life, and interrogating this belief is foundational to the kinds of questions and search for answers that follow in the pages of this book.

Colonial powers have worked persistently to replace, through dominance and lethal violence, Indigenous ways of relating to other people, animals, and ecologies, normalizing their anthropocentric dominance. And in this normalization, these dominant narratives obscure the fact that there are other, better ways of understanding the world and the constellation of relationships in which we are embedded. Indigenous ways of knowing, for instance, typically do not organize the world in binary or hierarchical ways, oriented as they are instead through an ethos of symbiotic webs of existence where humans aim to be stewards and coinhabitants with other forms of life. Even Indigenous conceptions of the very basis of life disrupt colonial ideas of *what life is*. As Kim TallBear writes, "For many indigenous peoples, their nonhuman others may not be understood in even critical Western frameworks as *living*. 'Objects' and 'forces' such as stones, thunder, or stars are known within our ontologies to be sentient and knowing persons."[4]

In the most simplistic sense, a colonial hierarchy is organized with humans at the top—as central and exceptional—and nonhumans at the bottom. Nonhumans, as forms of life with the lowest moral value, anchor this hierarchy. Value is defined in large part by what it is *not*, and the *human* needs the *nonhuman* to solidify its definition. But this ordering is more complex than a simple human/nonhuman binary. It is also a racial hierarchy ordered along socially constructed lines of difference *and differential value attached to these differences*.[5] Race and other perceived sites of difference, like gender, ability, sexuality, socioeconomic status, and geography, are categories that solidify what is perceived to be a *full human*. The *anthro* in *anthropocentrism*, then, refers to a narrowly conceived kind of human, and this identity is crafted as exceptional, holding the highest moral value. Racialized people, like Indigenous communities, are thus rendered subhuman or less-than-human to bar them from membership in this conception of *the human* and legitimate their exploitation and annihilation.

Capitalism thrives on, and relies on, this system of differential moral value. The construction of who is cast as the "other" to the *full human* shifts amorphously based on the need for an exploitable population to serve capital accumulation. Although lives rendered subhuman, less-than-human, or nonhuman are crafted by colonial ordering as having lower moral value, they have significant *economic* value as laborers, producers of commodities, and commodities themselves.[6]

For humans, antebellum slavery crystallized what it meant for humans to be ownable and wholly exploitable to drive capitalist growth, both to produce globally traded commodities (like tobacco, cotton, and sugar) and to be traded as commodified human lives. Today, these forms of extreme human exploitation manifest in ongoing forms of global slavery and debt bondage, as well as in, for instance, prison labor, child trafficking and labor, sweatshop labor, and migrant farmwork, creating populations of workers whose lives are subsumed by capitalist logics of commodified property and sustained fundamentally by a persistent colonial ordering of life.

Animals, anchored at the bottom of a colonial hierarchy, are overwhelmingly conceptualized as property, enabling the seamless appropriation of their lives and deaths as commodities and producers of commodities under capitalism. This property status confers on them categorizations as pets, as farmed animals, as biomedical research subjects, and as forced entertainers, and, in the context of wild animals, for instance, as objects of tourism, like the orcas, or as state-managed populations that may be sentenced to death through population culling or through the seasonal sale of permits to hunters. Across these designations, there is significant variation within the category of *the animal* that dictates how animals are assigned both economic and moral value: how they are exploited and/or reviled, and how they are protected and/or revered.

Elizabeth McKinley and Linda Tuhiwai Smith explain that settler colonialism is an *organizing structure* for society—not confined to a specific period in the past but enduring into the present and touching every part of contemporary life in settler states like the United States and Canada.[7] Capitalism, too, is an organizing structure. Each is unique, with distinct histories and contemporary impacts, and they work in concert to cement extractive and alienated ways of thinking about other beings and the ecological systems with whom we're in relation. The commodification of life is only possible because notions of private property, and the foundational forms of objectification that this designation entails, exist in the first place. These systems thrive on these dynamics of property, of objectification, of commodification, and of profit as core organizing logics, social structures, and political-economic systems today.

We're all of us—humans, animals, and ecologies—haunted by these systems and the relationships of care and sensitivity they foreclose. We're all, in fact, harmed by these structures no matter what position we occupy, as they foreclose possibilities for more caring, expansive ways of understanding who we are at our most fundamental essence and how our lives are entangled with the essence of others. They are so normalized, so very much just the way things are, that to imagine worlds otherwise might seem to be an impossibility. They shape our most mundane encounters with other species, and to transform these ways of understanding and relating to others, it's necessary to look carefully at these structures in which we're embedded, as well as the micro-moments in life that are affected by them. This can be a window into a different future—a more fulfilling one, one where we're more deeply connected to each other and to the meaning that comes from this connection.

In my own life, I've felt a sense of being haunted by the constraints of what feels impossible within these enduring systems. It's a feeling of unease, like what I felt stepping off the whale-watching boat that day. It's a shadow, a void of possibility. I've felt a longing for something that's mysterious, something out of reach, something I don't know enough to know. But there's an intuitive tug toward that *something*—a pull toward a glimpse of what might have come before all of this and a longing for what might come after it.

In *How to Do Nothing*, Jenny Odell tells the story of a pair of black-crowned night herons who perch in front of a Kentucky Fried Chicken (KFC) fast-food restaurant night after night in her neighborhood in Oakland, California. "The KFC," she writes, "is near Lake Merritt, a man-made lake in a completely developed area that, like much of the East Bay and the Peninsula, used to be the type of wetlands that herons and other shorebirds love. Night herons have existed here since before Oakland was a city, holdovers from that marshier time. Knowing this made the KFC night herons begin to seem like ghosts."[8] These ghosts—these living specters—persist, crafting their lives in the midst of and in spite of the demolition and attempted erasure of their ancestral lands. Their homes have been occupied and cemented over, refashioned into human-oriented urban environments. The night herons live as witnesses to what has otherwise been erased. They stand there as ghosts of their ancestors whose presence lingers somewhere in their DNA, their bones, their sinews, the sharp knowing gaze with which they survey passersby.

We might think about ghosts as figures out of place—as figures who linger, who refuse to leave a place, even when what is rendering them out of place insists on their annihilation. Ghosts are often conceived of as those who are dead and gone but who remain in some ethereal sense to haunt us. But ghosts might also be understood as those living who remain, who have been rendered out of place and who have persisted. The herons stand there as if to say, *Here we are. We're still here. We're still alive. And we see what you've done.* The harm against these herons, if we attend to it, is reflected back at us. Their out-of-placeness at the KFC can illuminate for us *why* they are out of place—not because they've come from elsewhere and appeared here, by chance, in this unlikely concrete habitat, but because they have been *made* out of place. Humans have built cement architecture up around them, making their flourishing all but impossible. And still, they persist. Those who refuse to disappear, to be displaced, and to die haunt that which has attempted to erase them. They are there, a haunting, but also living beings—refusing to be disappeared, and in this refusal, insisting on life, on liveliness, on standing outside that KFC and bearing witness through their presence to the destruction that has rendered them ghosts. To bear witness to them is to acknowledge the violence that has occurred and to be present for their ways of living in the face of it.

Avery Gordon writes: "What [the ghost] represents is usually a loss, sometimes of life, sometimes of a path not taken."[9] The night herons embody a profound ancestral loss; they stand there, witness to the loss of particular, black-crowned night heron lifeways. They are also there as a reminder of the loss of "a path not taken." Whose path? Certainly, their own but also a path not taken by humans who, in the past or the present, could have veered sharply away from the forms of harm and erasure that led to these forms of loss in the first place. The presence of ghosts, as the figures who remain, is part reminder of what has happened, part glimpse into what could have been.

These movements back and forth in time and the structural constraints of what kinds of lives are possible I see as forms of haunting. Eve Tuck and C. Ree describe haunting as "the relentless remembering and reminding that will not be appeased by settler society's assurances of innocence and reconciliation. Haunting is both acute and general; individuals are haunted, but so are societies."[10] Attentiveness takes the form of working to understand the broader structural hauntings of contemporary life, its careful nuances, and the singular individual encounters of daily life. Attending to haunting means attending to ghosts like the pair of night herons. It means attending to the structures that have rendered them ghosts, and it means attending to these two individuals—to their experiences, their lifeworlds, their practices of kinship.

"The ghost," writes Gordon, "imports a charged strangeness into the place or sphere it is haunting."[11] I read this as ghosts "making the familiar strange"—an effort to look anew (or askew) at that which feels so familiar as to not, perhaps, be felt at all. Attending to ghosts is a practice, in part, of injecting feeling into our forms of attention—of pushing back against the kinds of alienation embedded in refusing to acknowledge the ways we are haunted. "From a certain vantage point," she continues, "the ghost also simultaneously represents a future possibility, a hope."[12] Reckoning with ghosts, with haunting, is a matter of reckoning not only with violent pasts and presents but also with the kinds of futures that are to come.

In this context, I wonder over a series of questions: How might we understand what living means in a dying world? How might we think about the role of the witness, the ghost, as the world around us crumbles and, perhaps despite our best efforts not to, we are the cause of this crumbling? And who is this "we"? What can hauntings from the past and hauntings from a dark future-to-come tell us about our place in a world shared with so many others—those who are flourishing and those who are not? How do we reckon with the now-widespread knowledge that much of the human species has, in such a wholesale way, made the world uninhabitable for so many species, including our own?

Through close attention to the tiny details of everyday life in our encounters with other species we discover ways to think through these questions not with grandiose reflections about extraordinary acts of violence or care but through attention to the forms of violence that are so mundane they are hard to recognize as violence. Attention to the acts of care that might feel like so little but that, in fact, matter quite a lot. To the ways that these simple practices of attentiveness might become profound rituals of resistance and meaning.

The taken-for-grantedness of capitalism and settler colonialism as organizing structures is in part sustained by *inattentiveness*. When I think about the black-crowned night herons at the KFC, I think of the kinds of attentiveness that are obscured in capitalism and those that are made possible through intentional forms of witnessing these manifestations of harm. That the herons' ancestral home had been replaced with a fast-food restaurant seems to me a perfect reflection of the precise kinds of harm that capitalism as a structure has wrought. We might think of fast food as a pinnacle of US American consumerism and consumption—consumption that can happen so fast you barely notice that you've done it. Grabbing some chicken nuggets and fries from the drive-through to eat distractedly on your way from point A to point B is, perhaps, the opposite of attentiveness. My intention here is not to criticize those of us who eat fast food; it's more just to point out the logic that underwrites this consumption.

Consumption happens more quickly and abundantly when there's a lack of attentiveness, and thus capital accumulates rapidly precisely because of this *in*attentiveness. In turn, capitalist accumulation also flourishes in this void of inattention because inattentiveness obscures the conditions that are beneath that consumption. These are harms—like the uneven subsidization of food that makes fast food cheap and widely available for people experiencing food insecurity—that if they were attended to might be cause for pause, and which might lead to a grinding halt in the gears of capitalist production and consumption. Perhaps this halt might make way for the conditions where food is stripped of its commodity status and where a more equitable system of growing, sharing, and eating nourishing and culturally appropriate food might be possible.

Attentiveness in the face of structures whose very maintenance thrives on inattentiveness is a way of pushing back against these ingrained forms of harm. We might think of attentiveness as a politicized kind of witnessing. This witnessing might be a way to drive a tiny crack into this persistent architecture, and when this crack emerges, it might be possible to see that these structures were not as solid as they seemed but that instead they are already crumbling under the pressure of trying to withstand the violence they contain. Exploring these entangled structures illuminates insights about the effects of capitalist and settler colonial logics and processes on our shared multispecies worlds—dying worlds—and reflections on how different futures might manifest after and outside of these structures.

As I've thought about the places that I inhabit, going back to childhood, and how I might practice a greater sense of attentiveness, I've noticed that it's easy for me to imagine that I belong where I am. That I am entitled to this place, this land, my early morning walks. I feel at home here. I find meaning in everyday kinds of landmarks and encounters here, the place itself. To be attentive means, I think, first attending to the taken-for-granted ways that I (and all of us) relate to shared places, to relationships with others, the places and kin that make up a sense of home.

The way we are situated in the world shapes how we inhabit space, how we treat others, how we form unique ways of knowing. So, first, what does it mean to be situated? As Deborah Bird Rose has explained, "Our situatedness involves both our membership in the species that is responsible for so much harm, and our embodied, emplaced existence within the social and ecological domains of our lives. To be situated requires us to have knowledge of our place within our ecological contexts, and this requirement poses a problem for us because so much of the harm happens either at a distance from us, or in contexts that we are not well trained to see and understand."[13]

It's a struggle to attend to the harm caused so routinely to other humans and animals, to ecosystems, and to each other that we might not be able to see it as harm. The beliefs that organize how humans are positioned in the world have organized multispecies relationships in profound ways. They have dictated who is ownable and owned; whose bodies will be subjected to sexual violence, medical experimentation, and forced labor; whose premature deaths are such a routine part of everyday life that they are hard to even notice. Interrogating what it means to be human involves asking what kinds of harms are embedded in embodying this category and what kinds of possibilities for new ways of thinking this consideration of harm might open up. Put another way, might we consider that it's not *that* we are human that causes harm but *how* we inhabit what it means to be human?

We're living in a time when reflecting on how humans will live on this earth into the future is more necessary than ever before against the massive scale of environmental collapse and global crises wrought by human economic systems, politics, and an ethos of excess. Thinking critically about our own culpability in harm at a global scale is intertwined with the urgency of the climate crisis—setting a backdrop of fear and often hopelessness. As the Michi Saagiig Nishnaabeg scholar Leanne Betasamosake Simpson explains, "Focusing on imminent ecological collapse . . . is so overwhelming and traumatic to think about, that perhaps people shut down to cope. That's why clearly articulated visions of alternatives are so important."[14] We need an antidote to overwhelm. Looking clearly and humbly at the harm in the micro-moments of our everyday lives—reckoning with it in meaningful ways—so that we might imagine and then *live* these alternative visions can be this antidote.

In the stories and reflections throughout the book, I often write from and through the language of "we."[15] I want to take a moment to think through who is this *we*? Sometimes, the *we* means humans as a species (*Homo sapiens*), and sometimes it's a *we* of white settler descendants. Other times it's the *we* of earth beings, human and nonhuman. My intention is not to exclude or erase differences that constitute different notions of the *we*. The use of *we* can be a project of exclusion for those who are not part of a particular *we*. It can also be a project of homogenizing and collapsing meaningful differences—an effort at assimilation through language that claims inclusiveness. It can be a disciplining, an erasure. I want to highlight the shifting *we*s that might help to make sense of the place we each inhabit in the world, the kinds of relationships we find meaningful, the kinds of futures that are possible. In that spirit, I invite readers to consider the *we* as it manifests throughout these pages and to inhabit each *we* as you read only if it feels like it fits.

For me, when I was writing this book, home was Seattle, the ancestral lands of the Coast Salish, specifically the Duwamish Tribe, and the native home of countless nonhuman animals and rich ecological systems over millennia—some flourishing, some extinct, and some hanging on at the edge.

Geographically, Seattle sits in the Puget Sound region on Elliott Bay, in western Washington state, in the Pacific Northwest of the United States. It is a part of the world that is usually lush and green, with light rains and mist in the winter and clear, sunny days in the summer. The region is changing, though, as the climate crisis drives the weather to extremes, here and everywhere. The summers have gotten hotter and drier, and heat waves kill residents in the region on the hottest days. With the heat and drought comes the smoke from dozens of West Coast wildfires, making the usually crisp and clear air hard to breathe. Summer used to be what Seattleites referred to as the region's best-kept secret (the secret that summers were perfect—sunny, not too hot or too humid—and not at all embodying the rainy reputation that people who don't know the region might imagine). Now, summer brings conditions that make many of us long for autumn to arrive, for the relief of rain, cloudy gray days, and a breath of fresh, moist air.

The autumn of 2021, when the weather finally, blissfully, turned to fall and the rains came, I could feel the region taking a deep sigh of relief. But the rains came hard. Normally light and misty, that year they were torrential, and parts of the region flooded with the sudden influx of water. There I was, writing this chapter as the rain pounded down on the roof, a constant *drip drip drip* of water leaking through the window seal and making a fast-growing puddle on the inside of the sill. I got up to put a towel there to catch the drips. The news told us about the category 5 "atmospheric river" hitting the West Coast, delivering a water volume fifteen times greater than the volume of the Mississippi River. It seemed like every day I was learning new terminology for a weather phenomenon I didn't know existed.

Living, today, is an experience of each year transforming what we think we know about the world, about weather, about the changing climate, and about the places we know well. The world, perhaps, feels less recognizable with every year that passes. As I sat listening to the *drip drip drip* of the rain, I thought about how our sense of home is less recognizable than it was. We lament: *This isn't how Seattle is supposed to be. This is like the heat of Tampa. This is like the drought of Tucson. This is like the air quality of Pittsburgh. This is like the torrential rains in New Orleans. This is like. This is like. This is like. Anywhere but here.*

This feeling has been called *solastalgia*—a form of nostalgia for the familiarity of one's home disappearing in the face of profound environmental change. *Solastalgia* combines the word *solace* with *nostalgia*, a longing for solace. As Glenn Albrecht and his coauthors explain:

> The word "solace" relates to both psychological and physical contexts. One meaning refers to the comfort one is given in difficult times (consolation), while another refers to that which gives comfort or strength

to a person. A person or a landscape might give solace, strength, or support to other people. Special environments might provide solace in ways that other places cannot. Therefore, solastalgia refers to the pain or distress caused by the loss of, or inability to derive, solace connected to the negatively perceived state of one's home environment. Solastalgia exists when there is the lived experience of the physical desolation of home.[16]

The multispecies relationships that feature in the stories I share with you are taking place against the backdrop of a catastrophically changing planet, and this shapes the kinds of observations, relationships, and futures that are possible to imagine. Attending to this context and to solastalgia can be helpful, I think, in understanding other related forms of connection and alienation in multispecies practices of kinship and homemaking. The creation of place—of home—is a relational process; that is, places are shaped by the relationships that unfold in them, creating the contours of the feeling of a place, the landscape, the kinds of life that can flourish there, the forms of care that these relationships necessitate.

Through attending to place and, in particular, the multispecies co-creation of place, I find myself sometimes feeling a sense of nostalgia and seeking solace from forms of connection with others that feel like they're part of the past, but of which I have no tangible memory or experience. Solastalgia is its own kind of haunting—yearning for something that's gone, for something we might try to grasp as it slips away, or something we might try to reach back into the past and pull forward. But the past might not contain even the traces of what it is we're truly longing for because we've not experienced it yet. Or might it?

Solastalgia is also a kind of homesickness. But what does it mean to feel homesick for a home or set of relations built on layers of violence, for a home whose very construction and maintenance were erected on the annihilation of others who also inhabited and continue to try to inhabit this place as home, and which has delivered us to our current state of crisis? How does our need for solace intersect with or affect other animals, and their notions of homesickness for a past that is gone and a future that holds little hope? What kinds of alienation are built into the very notion of home—a refusal or inability to confront the kinds of truths about history, about ourselves, about the way we live on the earth that foreclose a more meaningful sense of connection to multispecies others and to places themselves?

Might we think about solastalgia not only as a longing for a past that is gone but as a hope for the future, a longing for different ways of relating to others, of more deeply connected notions of home and kin?

Several years before our whale-watching excursion, an orca in the Pacific Northwest made international news. On July 24, 2018, the orca Tahlequah, also known as J35, gave birth to a calf, Ti-Tahlequah, off the coast of Victoria, British Columbia. Shortly after birth, the calf died. Following her death, Tahlequah carried Ti-Tahlequah's remains on her nose or by one fin through the Salish Sea for seventeen days in a journey of grief. Each time the calf fell from her grasp, she dove down deep underwater to retrieve her infant and continue her journey again. As she swam—a thousand miles by the time the remains of the calf finally sank into the deep for the last time—other members of her pod gathered, swimming alongside her, bringing her food, and keeping vigil as she and they mourned.

Hearing this story affected people around the world, reflecting empathy for what many recognized as a profound form of grief and prompting reflection on the inner lives of other animals. How might we understand Tahlequah's grief and the collective mourning of her clan? We might ask, what more is behind Tahlequah's publicly recognized process of grief. What are the structural and anthropogenic contexts for this loss? What kinds of relationships with others in her pod and clan help to sustain her and create a vital kinship network through this loss and through intergenerational losses? What are the things we will never be able to know—the questions we haven't even thought to ask?

Tahlequah is a member of the J pod, one of three pods (J, K, and L) that make up the endangered Southern Resident clan. Reasons for their decline are multiple and are driven by the effects of capitalism, including toxic exposure to water pollution from industry and the numerous impacts of shipping vessels on their habitat. At the heart of their decline is the scarcity of their primary food source, the Chinook salmon. Without the Chinook, the Southern Residents are more vulnerable to the other threats to their well-being. For instance, when they are not able to find enough to eat, their bodies start to metabolize their fat, drawing into their bloodstream toxins that have accumulated in their blubber. The Chinook salmon populations themselves are in a dire state—in Puget Sound, they are at only 10 percent of their historic numbers.[17] Scarcity of Chinook salmon correlates with high rates of miscarriage and reproductive problems among the Southern Residents who rely on them for food; in the Southern Resident population today, 69 percent of pregnancies result in miscarriage.[18] Like those for the Southern Residents' decline, the reasons for the Chinook salmon's dwindling numbers are complex, but chief among them are overfishing by humans and the changing environment due to the anthropogenic climate crisis. The irrevocable harm that humans have caused haunts the Southern Residents and the very possibility of their survival. The birth of Tahlequah's baby was consequently met with great anticipation and relief as the Southern Residents hadn't had a calf in quite some time.

Orcas form matriarchal cultures, where knowledge is passed down matrilineally, especially by those females who have reached middle age and gone through menopause. Matriarchs are central to the health and survival of orca pods; they are leaders, helping to find food, and forgoing food themselves when there is scarcity, even if it means sacrificing their own health and well-being.[19] In 2019, Tahlequah's mother, a forty-two-year-old, middle-aged matriarch in the J pod, called Princess Angeline or J17, was presumed dead after her condition had worsened in the preceding few years.[20] Researchers had been worried about the marked wasting in aerial photos of the orca, a sign of inadequate food sources and severe malnutrition, and as they had feared, she had likely died as a result. In the year following the death of Ti-Tahlequah, the death of this matriarch marked the deepening loss of vital ancestral knowledge about Southern Resident lifeways, and these damages reverberate through the clan across generations.

This harm is felt in the afterlife of the intergenerational trauma the Southern Residents suffered as a result of the State of Washington allowing for the widespread capture of orcas for marine parks in the 1960s and 1970s. In what is referred to as the Penn Cove roundup of August 1970, seven Southern Residents were captured, made into property, and transported into lives of captivity and isolation in marine parks like SeaWorld.[21] During this roundup, four other orcas who would not abandon their kin drowned in the nets and then were surreptitiously disposed of so as not to have their numbers documented as part of the allowable limits of capture.[22] This abduction and killing of Southern Residents created a significant rupture in the social structure and kinship relations of the clan, which in subsequent decades has been difficult to repair with the added pressures of anthropogenic effects on orca habitat, food sources, and health, and the resulting reproductive difficulties that have also contributed to the population's decline.

It is in this historical context that Tahlequah's loss resonated with millions of people, echoing the catastrophic losses we have all experienced, the deep wells of grief that many of us hold within us. Orca researchers have found that orca brains are incredibly complex and that, for instance, the part of the brain that controls emotion is significantly more developed in orcas than in humans.[23] We might ask, then, what this means in terms of our ability or inability to understand the depth of orcas' emotions—the depth of Tahlequah's grief. Deborah Giles, a research scientist with the University of Washington's Center for Conservation Biology, in an interview with the *Seattle Times*, said of Tahlequah as she swam carrying her calf day after day, "What is beyond grief? I don't even know what the word for that is, but that is where she is."[24] It's humbling to not be able to name or understand a loss so profound it is beyond our conception of grief. It's humbling to realize and attend to the fact that there is so much about other species that we might never be able to understand or the depth of feeling we might never experience.

I've chosen to share this story because I think it is beautiful and devastating and a symbol of the perhaps irreparable ways that humans have harmed others in our shared world. But I also wanted to share this story because many people readily acknowledge and take seriously the complex inner lives of orcas and other cetaceans. These species fall into the category of *charismatic megafauna*—those species who loom larger-than-life in some way for humans, perhaps because humans see themselves reflected back to them in all of their behavioral, emotional, and intellectual complexity. Orcas are an easy access point for humans who may be skeptical about the cognitive worlds of other animals to acknowledge that *these animals* might be worth understanding and caring about. We think we can understand orcas because we recognize something of ourselves in them. But do we understand them? Can we? And do we need to?

Two years after Tahlequah lost her calf, on September 5, 2020, she gave birth to another calf, known by humans as Phoenix, or J57. Just in advance of his birth, all the Southern Residents (J, K, and L pods) gathered in the Salish Sea for what we might understand as a greeting ceremony, a celebration of the new birth. Some had come from a hundred miles away to arrive in time for the birth, and Phoenix was born into the clan amid the excited whistles and clicks of his orca kin. How they knew to gather on this day, to forecast the birth across a distance to multiple pods, is a mystery to researchers. "They seem to have some kind of communication system that's hard to imagine, because they were 100 miles apart and around several islands, so out of acoustic range. But somehow they were able to meet in that location at that time on the day of the birth of the new baby," Howard

Garrett, cofounder of the Orca Network, told NPR.[25] After the collective ocean of grief left by the death of Tahlequah's calf two years earlier, the euphoria of the celebration of life and the continuation of the lineage of the Southern Resident orcas was palpable, crossing species lines and moving many humans who had never seen or met this family of cetaceans.

I explain this not to privilege orcas and their experiences over other animals but just the opposite—to try to disrupt humans' hierarchical ordering of species that positions humans as exceptional and superior, as the pinnacle of evolution against which other species are measured. Glimpsing the possibilities that orcas feel things more deeply and process things in much more complex ways than humans could have the effect of merely accepting certain other species with what humans designate as more cognitively complex lifeworlds (e.g., whales, elephants, African grey parrots) into a circle of greater recognition and respect. However, I'd like to suggest that this might be a window into a different way of thinking—a thinking that acknowledges that there is much we don't know about how other animals experience the world and that human experience is not the only or the most valuable way of inhabiting the world. The human brain and psyche are perhaps not a measure against which other species should be compared. This not only addresses the problem of an anthropocentric worldview but also offers insight into the issue of anthropomorphism, which is worth taking a moment to explore.

Anthropomorphism refers to the practice of attributing human qualities to other species, and to the process of trying to understand a member of another species through a human lens. It's a slippery concept, and there are a few different ways we might think about it. To a certain extent there is no way to avoid anthropomorphism from the standpoint of being human. We will never, no matter how hard we try, be able to truly divorce our way of understanding and seeing other species from our perspective as humans. Studying the lives of other species is often a difficult task. Our ways of knowing are filtered through our own human lens, and through our own social and cultural positioning in the world. There is, then, a certain level of anthropomorphism that is inevitable—it is a matter of simply using our human viewpoint to identify and name what we see in other species.[26]

The attribution of what many humans might define as uniquely human characteristics to nonhuman animals emerges as one critique of anthropomorphism, rooted in the belief that humans are the only ones to experience complex emotions or a sense of self. However, research is rapidly accumulating to show that many different species possess rich emotional and cognitive inner worlds and that there may in fact be very few characteristics that are truly uniquely human. So, there's a way that lodging the accusation of anthropomorphism is anchored by an anthropocentric worldview, wherein a person's imagination is limited to believing that humans are exceptional and thus superior to all other animals, or at least *most* other animals.

There is another serious problem with anthropomorphism, I think, which is that the anthropocentrism at the heart of anthropomorphism precludes seeing the ways that other species are unique. In essence, anthropomorphizing an animal (making them seem humanlike) strips them of *their* animality—the qualities that make them uniquely an orca or uniquely a rat. We see this with the example of orcas' emotional capacities being more highly developed than humans'—that if we were to limit our imagination to projecting our human understanding of grief onto orcas, we would miss a consideration of how orcas may feel grief differently or more deeply than we can even comprehend.

Anthropomorphism is also anchored by a tendency to make broad generalizations about the characteristics and capacities of entire species, ignoring the way that, just as in the human species, there are dramatic differences among individuals, and different communities in different places create unique cultures. So, I want to take care to avoid making claims that *orcas as a species are like this* or *cows are like that.*

Even more problematic is the delineation between humans and all other animals, which ignores the fact that humans themselves are animals and simultaneously homogenizes all nonhuman species into one group. At the level of species, there are certainly some observations we can make, such as the fact that orcas have some particularly complex parts of their brains. But how this manifests for individuals will vary widely, as it would in humans. In essence, Tahlequah might experience grief in a very different way from another orca in her pod, and this may be expressed in different kinds of rituals. Not all orcas, for instance, carry their dead calves a thousand miles in a journey of processing a catastrophic loss. There are many other ways of processing grief—some of which we might not be able to recognize or imagine.

I think it's important to try to understand, to try to make sense of our impact on those around us, but I also want to propose that perhaps a full understanding isn't possible. As pattrice jones explains, "We have a hard time even imagining how awareness or intention might be realized in beings very different from ourselves. That's okay. We don't need to know or even be able to imagine everything. As long as we recognize that there are limits to our knowledge and imagination and that things may be true even though we cannot perceive or prove them, then we can avoid the folly of unfounded assumptions loosely rooted in ignorance."[27]

As our whale-watching boat headed back to shore that day, many of the passengers sat and rested inside the boat's cabin, enjoying the company of those they were with, and eating snacks from the canteen on board. We approached the naturalist and told him that we'd heard the stories of Tahlequah in the news years before—the loss of her baby and the jubilant gathering around the birth of her next, Phoenix. His eyes glimmered, and he said quietly, "I was there at the birth of Phoenix." He explained that he'd been out kayaking that afternoon and suddenly the pods of orcas gathered not far from where he was paddling. He watched as they celebrated this new birth in the setting sun, stunned to be receiving this gift—just one human in a tiny boat observing in silence something so much bigger than himself.

Although I didn't ask him, I wondered why he didn't talk about this remarkable experience in his presentation. I wondered if it was because there was something truer, more meaningful, about his being there in his kayak, silently observing this wholly *orca* event. I wondered if somehow integrating it into daily public presentations felt like it violated the experience. It was an experience so unlike that of the whale-watching boat, so completely unexpected and unanticipated, something so humbling in its magnitude, that perhaps there weren't words that adequately described its magic.

Attentiveness requires this kind of humility and an awareness of the limits of our capacity to understand, to know, to seek new ways of knowing, to fail, and to accept the kinds of mysteries—the secret lives of others—that are out of reach.

2

Journeys Ended Here

The transformation of the heart such beauty engenders is
not enough tonight to let me shed the heavier memory,
a catalog too morbid to write out, too vivid to ignore.
The weight I wish to fall I cannot fathom, a sorrow over the
world's dark hunger.
—Barry Lopez and Robin Eschner, *Apologia*

The fox was a flash, a ghostly figure darting across the highway just ahead.
It was a foggy night on a dark Vermont highway, and at first I thought I
was seeing things: shapes, bodies, and lives in the fog that weren't really
there. But as the car approached, the fox was visible more clearly for a moment before disappearing into the forest—safe. I breathed a deep sigh of
relief and whispered a quiet *oh, thank god*.

On the drive up to Vermont from Connecticut to visit friends, I passed
so many animals—or rather, their remains—on the side of the road that
I lost count. The farther north I traveled, the more densely the shoulder
of the road was populated by corpses. Some were identifiable, like the
mounded bodies of raccoons, their unmistakable striped tails visible
as I whizzed by. The contorted bodies of deer sprawled out on the pavement, dragged unceremoniously and, no doubt, with haste, to the side

of the road to prevent further collisions. A red fox, lying on their side, looked like they were sleeping. I wondered whether someone had struck the fox and moved their body, or whether the fox had been hit and then made it on their own out of the roadway to die just out of reach of the passing traffic. These were animals I could recognize as I traveled at sixty miles per hour toward my destination. Others were unrecognizable. These were bodies so disfigured by the violence of a motor vehicle collision that they looked like only a tumble of matted, bloodied gray or brown fur. A squirrel? Groundhog? Possum? Rabbit? Before I had a chance to think too much about it, though, I was passing someone else's remains.

These were all the visible animals—ones who had not yet been picked up by state agencies for rendering, composting, or delivery to the landfill, and those who were large enough to see from a fast-moving vehicle. The millions of insects, amphibians, and small birds who are killed by cars are rarely seen or registered enough to be included in this accounting from the driver's seat. Sometimes, when I travel on roadways, a death, a body is only registerable through scent—the sharp smell of a skunk, or the nauseating smell of a decomposing corpse in the heat of the summer sun.

Occasionally, I see a bird of prey—a hawk or a turkey vulture twisted on the side of the road, a wing angled up at the sky, feathers shifting in the wind of passing traffic. I wonder if they'd been hit while scavenging the body of another animal killed by a car or truck. I've seen turkey vultures and hawks picking at the bodies of dead animals on the road ahead, or on the shoulder, as I pass. They could easily be struck in these moments. Or they could be hit as they are hunting for not-already-dead food. I imagine a hawk swooping down across the highway, eyeing a small bird or rodent, a semitruck plowing into them, crushing their body midflight.

The experience of witnessing animals who have been killed on roads is overwhelmingly an experience of *bodies*. It's not an experience of encountering a living animal moving through the world, on their own journey— the magic of seeing a fox running swiftly and quietly through the foggy night. Instead, all we can know is what we can witness about the body—a twisted, contorted, bloody figure rendered dead by human movement. Our own bodies, too, respond—alive—a visceral punch in the gut at the sight, a wave of nausea (disgust mixed with horror mixed with grief?). What can we know through attention to a body, dead? We are left to wonder.

"Who are these animals, their lights gone out? What journeys have fallen apart here?" Barry Lopez asks of animals killed on roads in his book *Apologia*.[1] Each time I read Lopez's words, I feel a well of emotion, of grief not merely in his acknowledgment of the fact of the animal dead on the road but in his call to wonder, to care about these animals' lives and what was lost, for them and others, with their death. I'm left with more questions.

What had the world lost—that we didn't even know it had lost—when this animal died? Who had these animals left behind? Were their kin waiting in the forest somewhere for them to return, or were they looking on, just out of sight beyond the road, watching as their loved one was killed and lay dying or dead in the road? What, for instance, of a doe reported dead on a road in North Dakota, leaving twin fawns who stayed with the body of their mother for four days, until her remains were picked up for disposal? What would become of them, those fawns left behind?

It's rare to think deeply about animals killed on roads. Some people don't think of them at all. For others, it's painful to think too much about these deaths—about the pain or suffering of these animals who've died, and to think of who might be grieving them from the forest or field. It's perhaps especially painful to think of them if we are the ones immediately responsible for their deaths. However, the nature of traveling by car at high speeds has a way of making us quickly forget the dead we've passed by—and maybe we feel thankful for the glossing over that might allow this forgetting to occur. It's been my experience that most people don't like to linger too long on uncomfortable truths about human violence against animals, especially when they've been complicit in their deaths, but also in accidental deaths, when there is no clear or convenient way to remedy the situation. It is easier, more comfortable, to ignore or forget the one million vertebrate animals killed *every day* on US roads (and the countless other invertebrates who are not counted at all).[2] But how many animals are not killed on impact? How many lie dying for hours or days? How many limp into the forest or field as far as they can manage? How many, then, are not counted at all?

Lopez writes, "We treat the attrition of lives on the road like the attrition of lives in war: horrifying, unavoidable, justified. Accepting the slaughter leaves people momentarily fractious, embarrassed."[3] And so, we immediately work hard to forget. To zoom by, leaving that feeling and the body behind. But I, and maybe others too, feel culpable in these moments. I feel shame in speeding by, in not stopping. In working to forget. I feel shame in actively participating in a car culture that is the cause of so much killing of those who do not in any way choose to enter into the violence of this culture. Shame can be an excruciating emotion—we feel a sense of responsibility for something that has gone wrong, we know our culpability, but we don't know, or perhaps won't act on, an attempt at remedy. It is painful, often intolerable, to sit with shame. Shame is an emotion of turmoil, an active struggle in the body and mind. And this is perhaps because, as Jonathan Safran Foer writes in his book *Eating Animals*, "shame is the work of memory against forgetting."[4] This struggle to forget what wants to be remembered works to more brightly illuminate the reality of how we are entangled with this violence. L. A. Watson writes, "The animal is hit, and we run—not only from our own culpability in the matter as individuals, but also from seeking to address the ways in which this violence and these deaths are systemically built into our own systems of survival."[5] What are these structures and histories that offer context to this killing and death, to this shame and forgetting? And how can attending to context help us to avoid a disavowal of these animals, our shared histories, and the possibility of a different way of relating to each other?

Animals killed on roads are characterized as *accidental deaths*—and to a certain extent, yes, collisions with animals on roads are accidents. Most people don't intend to hit an animal with their car. But the development of an extensive system of roadways that indiscriminately transect wildlife corridors, migration pathways, and habitats is hardly accidental. This development and construction was a purposeful and deliberate process and was undertaken for a whole host of reasons. At its root, road construction has been a key mechanism in building a more extensive capitalist economy throughout North America. The early construction of roads increased not only the circulation of goods as commodities but also the growth of automobility. The proliferation of car culture offered a site for incredible growth and capital accumulation through a manufactured demand for cars and through the sprawling development that automobile transportation allowed. It is impossible to understand the phenomenon of animals killed on roads without understanding the development of roads and of automobility.

I focus most intently on roads (and, secondarily, railways) since these are transportation networks where the deaths of animals are most visible. But animal deaths occur in other forms of transportation networks that are entangled with capitalism and colonialism, like on the water and in the air, where animals are killed routinely by boats and planes without anyone noticing or witnessing their deaths or their remains. The twin land-based transportation projects of roads and railways were instrumentalized to facilitate the murder and dispossession of Indigenous communities throughout western North America, the destruction of the prairies and forests to clear land for farming and ranching and create vast reserves of timber, and the extermination of native animal species, like bison and wolves. The expansion of settler colonialism as an organizing structure of violence in North America was, in large part, *made possible* by the development and expansion of both roadways and railroads.

As I researched the history of road construction and automobility in the United States for this chapter (a topic I never imagined exploring in any depth), I found myself both riveted by what I was learning and profoundly bored. I was riveted because I have an ongoing interest in understanding the history, background, and functioning of the most mundane features of everyday life, and roads certainly fall into that category. But at the same time, slogging through page after page of road history (of which it turns out there is a lot) was tedious, and I found my mind wandering constantly. I've learned over the years, though, that when I'm bored or when a subject makes my mind wander, I should look a little more closely and work a little harder to understand it. In the case of roads and transportation, the boredom that might arise in trying to learn about their histories could work to render these histories neutral or apolitical. (*Oh, they're boring; there must not be anything too dramatic or troubling happening here.*) Boredom dulls the senses, and it deadens the capacity to attend fully to a subject or experience. By contrast, taking care to pay attention to things otherwise ignored—activating the senses to experience the world around us in a heightened way—politicizes seemingly innocuous relationships that may in fact be constituted by violence.

I've tried to offer as concise a history as possible here, but if you find your mind wandering, your attention wavering and veering toward impatience, I invite you to pause and reflect not only on the subject at hand but on the very experience of boredom itself. What is the essence of that boredom? What subtle work is it doing to obscure the details of what might feel like an innocuous history? How can attending to that boredom transform in nuanced ways the nature of what you know and how you relate to yourself and the world around you?

The starting place for my research on roads was rooted in the experience of speeding along the highway in my car, witnessing the animals dead on the side of the road. But, quickly, this research catapulted me much further back than the widespread proliferation of cars. Road systems were an integral part of the settler colonial project in the United States and became tied to capital accumulation itself. Prior to the 1800s, roads were often constructed of unrefined dirt, gravel, or pounded stone and tended to follow topographical terrain. In the early 1800s, however, an extensive turnpike project was born, where private companies invested significant capital in the construction of toll roads throughout the Atlantic coast states, and especially between the large Eastern Seaboard cities (Boston, New York, Philadelphia, Charleston, etc.).[6] This turnpike system provided greater and more efficient connectivity, and it was not long before the Atlantic coast states sought the expansion of this system of roads westward over the Allegheny Mountains.[7]

It was at this same time that transportation systems, including decent roads, began to be viewed as integral to growing the "wealth of the nation."[8] In 1816, John C. Calhoun made a case for more comprehensive transportation systems: "If we look into the nature of wealth, we find that nothing can be more favorable to its growth than good roads and canals."[9] Roads and canals, then—and also the emerging railroad—became central to increasing the power and reach of the settler state westward, and one key way that this was achieved was through the increasing circulation of capital. In addition to providing more efficient modes of transport for goods that could now be sold across greater distances, thus growing the nation's wealth, a more cohesive transportation system increased land value. After the widespread genocide of Indigenous nations and the dispossession of surviving members of these communities of their land, the federal government claimed ownership over most of the so-called undeveloped land in the states. This land was more or less valuable based on its proximity to urban centers, and so federal support for a widespread transportation development project was in part based on the promise of increasing land values as a transportation network made these lands more accessible and thus desirable to private buyers.[10]

Calhoun also framed transportation as integral to the growth and strength of the *nation*: "Let us then bind the Republic together with a perfect system of roads and canals. Let us conquer space. It is thus the most distant parts of the Republic will be brought within a few days travel of the centre."[11] *Let us conquer space.* It was this sentiment, and a Manifest Destiny sense of entitlement, that drove settler expansion westward, along with the imperative to increase wealth—private wealth, the wealth of states, and the wealth of the federal government. Rooted in these colonial and capitalist logics, what and who the construction of roads, canals, and railways destroyed became collateral damage, as humans, animals, and the landscapes themselves were indiscriminately killed, displaced, and obliterated.

Ecologically devastating, the construction of the railroad involved cutting down trees (to clear space and to acquire timber for railroad ties), clearing brush, and blasting rock apart to cut through the landscape and level the path for the tracks to be laid. Bison, as a keystone species anchoring a constellation of other animal, plant, and human species, were systematically eliminated from the plains in part as a result of the proliferation of transportation infrastructure.[12] Bison were critical species for the health and flourishing of prairie ecosystems, but they were also an integral part of a "web of relations" involving close kinship bonds with Indigenous Plains nations. The loss of bison, though, was not only devastating for biodiversity and for humans. Their extermination meant that bison themselves suffered catastrophic losses of their own bison kin, a reverberating trauma through individual members of an entire species.

The subsequent construction of railways was performed primarily by Chinese immigrants, enslaved persons, "convict" laborers, free Black people, Indigenous persons, and European immigrants.[13] The growth of free-market capitalism in North America was and remains rooted in the genocide of Indigenous peoples and antebellum slavery, and capitalist expansion was enabled at least in part by the development of transportation networks—both in their construction and in the proliferation of their use. Contemporary roads and railways, then, are connected to this long and violent history of settler colonialism, antebellum slavery, immigration policy, and low-waged labor.[14]

The nineteenth-century frenzy of rail and road expansion shrunk time, accelerating the circulation of commodities and collapsing what were previously extraordinarily long distances via horse-drawn vehicles or boats for transport and travel. The historian William Cronon writes, "Nineteenth-century rhetoric might present the railroad network as 'natural,' but it was actually the most artificial transportation system yet constructed on land."[15] Whereas waterways and dirt roads had followed natural landscape features that promoted easier construction, the railroad cut across the landscape in the most direct routes possible, and this meant transforming the environment more thoroughly in its construction and maintenance.

With this land transformation, there was also a profound shift in people's expectations about time: that transportation should be fast, that they didn't need to wait long periods of time for the movement of people and goods, and that time could be divorced from the natural limits of seasonal travel and the life energy of individuals previously needed to transport themselves. Roads and railways shifted the way people moved across space, blurring the landscape as movement accelerated, allowing for its erasure and abstraction from what travelers were seeing as they passed by.

In the late nineteenth and early twentieth centuries, transportation development continued to build greater connectivity between places not already well connected. The result was an increasingly dense network of roads, rail lines, and waterways. With each new development came further incursion into the habitats of free-living animals and biodiverse ecological systems, cutting indiscriminately through the landscape.[16] One phase of this transformation of the land (although it accounts for only 1 percent of roadways in the United States today) came with the development of the interstate highway system in the 1950s.[17] The primary reasons for the development of interstate highways were economic growth, the alleviation of traffic congestion with the fast-growing number of motorists on the road, and greater safety. It was also at least in part meant to improve the ease of military movement and preparedness in the event of a nuclear attack on the United States.[18] The interstate highway project also, importantly, coincided with a time when the speed of motor vehicles was increasing, and when there were a growing number of cars on the road.

This proliferation of cars was enabled by Henry Ford's creation of the assembly-line model of automobile production in 1913, which led to cars becoming more affordable (and thus a more common consumer item) for the working and middle classes. The Fordist model of production involved a remarkable acceleration of commodity culture in terms of all kinds of consumer goods, such as food, clothing, household items, and cars. The car, suddenly, was both a desirable *and an attainable* commodity for the average consumer. The car became fetishized as a status symbol, a technology of convenience, a means for recreation. As I explore elsewhere in the book, violence and harm are central features of commodifying a living being (animal or human). But the production and circulation of *inanimate* commodities (like cars), too, have troubling impacts in their production, use, and circulation: in the extraction of raw materials for their production, in the labor involved, in their environmental impacts, and in their transport across space that enables consumption. The making of a commodity involves a transformation not only in how we view that thing but also in its material effects. Karl Marx's concept of *commodity fetishism* is useful here: it refers to the way commodifying a thing or living being obscures the origins of their production or existence outside of a capitalist economy.[19] As Dennis Soron argues, in fetishizing the car as a commodity, for instance, as a shiny new consumer item, or even as a necessary convenience, the car and its use become thoroughly normalized, and it is in part this twinned fetishization and normalization that obscure not only the reality of their production but also the harm they do in the world.[20]

As car use increased, rail declined, and this marked an important turning point in transportation history—investment in automobile transportation (a form of *individualized* transport) exponentially eclipsed public transportation systems (a form of *collective* transport).[21] This chasm deepened inequality in access to mobility. Whereas cars used to be reserved for the rich, now they were available to a much wider sector of the population, but there were still many people whose socioeconomic status did not allow them to own a vehicle. As car use became a norm, those without the means to purchase and maintain a vehicle were further limited in their mobility. Joseph Interrante explains, "What began as a vehicle to freedom soon became a necessity. Car movement became the basic form of travel in metropolitan consumer society. However, there was nothing inevitable about metropolitan spatial organization or people's uses of cars upon that landscape. The car could have remained a convenience used for recreation and cross-movement outside areas serviced by railroads and trolleys, while people continued to use mass transit for daily commutation. Car travel could have remained an option offering certain distinct advantages; instead, it became a prerequisite to survival."[22] This, in turn, enabled the dispersal of people away from cities and an increasing geographic division between the places where people lived and worked.

New forms of consumption were facilitated by the emergence of widespread car culture—for instance, homeownership in the suburbs, which allowed for larger, more expansive living spaces, thus creating a proliferation of consumption to furnish these new homes. Suburban retail industries cropped up to serve the suburbanizing population through the latter half of the twentieth century. The widespread consumption of cars as commodities, then, was an integral part of a growing and intensifying consumer culture that would deepen the multispecies harms caused by capitalism's need to constantly accumulate capital. Automobiles—and, indeed, growing automobile dependence—offered a clear path to achieve this aim, and so, without investment in robust (and less lucrative) public transportation systems, automobiles solidified their position as the primary mode of transportation for a growing and geographically diffused population.

Prior to the 1950s, animals killed on roads were typically limited to smaller creatures, like rabbits, raccoons, birds, and snakes. However, as the speed and number of automobiles on roads increased following World War II, mortalities of larger animals increased, bringing the problem of animals killed on roads into the public sphere, since these larger animals posed a threat to human life and property.[23] The subsequent transformation and exponential rise in automobility created the perfect conditions for vehicle collisions with wildlife, and since the 1950s, the number of cars on the road has continued to increase; vehicle registrations, for instance, doubled between 1970 and 2009.[24] To accommodate this dramatic increase in cars on the road, between 1980 and 2011, the construction of roads increased 6,500 miles per year, totaling 183,000 miles of new road construction in that thirty-year period.[25] As of 2011, the United States had a road network of roughly 4 million miles.[26]

According to the Federal Highway Administration, "Approximately one percent of all public roads are part of the Interstate Highway System. Of these 47,000 miles of interstates, 65 percent are in rural areas and 35 percent are in urban areas. Seventy-four percent of the remaining public roads are located in rural areas, with 26 percent in urban areas."[27] The rate at which crashes with animals occur is significantly higher on rural roads than on urban ones. The US population at the beginning of the twenty-first century was three times the population of the early twentieth century, precipitating the need for the expansion of both the reach of roads and the width (number of lanes) of highways.[28] Federal Highway Administration statistics for 2009 reveal that roads sustained three trillion vehicle miles traveled in that year, and 85 percent of these were traveled by private vehicles (cars, vans, motorcycles, SUVs, and pickup trucks), suggesting that much of the traffic moving on roadways is actually for personal use.[29]

Roads and transportation are profoundly mundane, rarely thought of in any detail by travelers unless something has disrupted smooth travel (a closed road, a bad traffic jam, a broken-down car, cavernous potholes, icy conditions). To reject this glossing over is to understand a different place of roads and automobiles in our history. The expansion of modern roads and increasing automobility, as Soron writes, "has led to an ongoing degradation and fragmentation of animal habitat, confining wild populations into enclosures too small for their needs and forcing animals to attempt road crossings for access to food, water, cover, migration routes, nesting sites and potential mates."[30] In a society where high-speed car travel is a taken-for-granted, unquestioned, and ubiquitous part of daily life, the devastation to nonhuman life becomes so normalized as to barely register as a significant problem. And this is despite the fact that humans kill more animals on roads than in any other sector of human activity after food production (which kills by far the highest number of both land and aquatic animals). Soron points out that "the automobile has been a leading agent of violence against diverse forms of animal life—including humans."[31] But the automobile is driven by a human person, it's made by humans and human-made machines, and it drives on roads constructed by humans who have prioritized human convenience and innovation over the well-being of other forms of life. The automobile becomes a weapon against nonhuman life, and this weapon is wielded by humans. *Humans*, then, via the automobile, have been "the leading agent of violence against diverse forms of animal life."[32] And acknowledging both the automobile's role and humans' culpability in so much violence lays bare the reach of anthropocentrism and the overwhelming harm it does.

I wasn't thinking about these issues at all, though, when, in 2006, my then-partner Eric and I took a round-the-country road trip. I had just finished college and had moved to Seattle; Eric was in between contracts for his job. It was the perfect time to take a trip like this. I didn't consider how a road trip primarily for recreation reflected a profound sense of entitlement—to moving through space without letting our impacts on the environment or on the animals we might kill while driving change the fact of our travels. It reflected the fundamental, taken-for-granted knowledge that the road was there *for us* to use however we'd like to use it, within the parameters of the law.

Originally, the plan had been to drive straight to Pittsburgh (my hometown), pick up some of the things I wanted to move out west, and drive straight back, a move westward that traced those same nineteenth-century pathways cut through the landscape by settlers and the settler state. Eric's stepfather had lent us his trusty old pickup truck for the trip, since we didn't have a car of our own. What was meant to be a two-week trip quickly started expanding as we planned our route. Ultimately, we took six weeks and first headed down the West Coast from Seattle, meandering along the Oregon and Northern California coast on the 101. We camped and wandered the beaches, took our time visiting the small fishing towns along the coast, and spent several days visiting Eric's brother and his girlfriend in Arcata, California. From there, we traveled south to Los Angeles, east through the Southwest and across to North Carolina's Blue

Ridge Mountains, north to Pittsburgh, and then west on the northern route, through North Dakota, Montana, Idaho, and back to Washington, visiting friends and family along the way, and camping before the weather turned for winter.

Road trips—especially ones that are thousands of miles long—are occasions to witness many hundreds, if not thousands, of dead animals on the sides of roads. The sheer scale of animal death over that many miles is almost too much to comprehend, and the individual animals blur together in an anonymous string of lives lost until they become, almost, just a mundane feature of the landscape. Is it possible to feel, intensely and intentionally, every dead individual witnessed on the side of the road? Even now, thinking back more than a decade later, I can remember only a few instances where individual animals we witnessed still register in my mind and heart. Part of this may be a problem with memory and forgetting— the intuitive process of actively working to forget things that are painful to remember and that remind me of my own participation in violent systems of travel and transport. The individuals who cut through that fog of forgetting were memorable for extraordinary reasons.

But this forgetting is also, at least in part, a problem of *paying attention* in the moment. To pay attention to every dead animal passed on the road is to divert attention away from other things; to pay attention to something else (what's on the radio, a conversation, the need for a rest stop, driving itself) steals attention away from animals on the side of the road. Perhaps, then, holding all of these things in the mind at once is an impossibility. Animals dead on the side of the road become easy to ignore, especially so, perhaps, because it is less painful to ignore them. It is much easier to focus on anything else, and thus refuse to give attention to the death and loss those anonymous-to-us nonhuman others are experiencing every day. It is, then, no wonder really that forgetting these animals feels natural and easy—we have only seen them peripherally, out of the corner of our eye, and soon they are out of our minds. In spite of this, though, there were some animals who haunt me—their stories, and their physical, embodied selves.

While we were visiting Eric's brother and his girlfriend in Arcata, I noticed a small table in their living room that was covered in treasures, a sort of altar of found objects from the natural world: there were pine cones; rocks, shells, and driftwood from the nearby beaches; pieces of bone found in the woods; and a chipmunk's skin and fur. I asked about the chipmunk and my brother-in-law went over, picked up the fur, and handed it to me. I stroked the soft, velvety fur, while they told us how that chipmunk's remains had come to be there. They had been driving on a curvy road that wound through the redwood forest, and this chipmunk darted out in front of their car so suddenly that they were unable to avoid hitting the small woodland creature. They were devastated. They pulled over and walked back to the site of the collision, and his girlfriend held the tiny lifeless body—still warm—in her hands. They discussed what to do. Confident that they didn't want this chipmunk's body to *go to waste*, they decided to take the remains home to eat. They skinned the chipmunk, cooked and ate the small, bare body, and then tanned the tiny fur coat that had recently enveloped and protected this once-lively creature. It was this chipmunk's remains that I held carefully in my hands. Even then, although I wasn't thinking about the politics or ethics of animals killed on roads in an intentional way, I wondered, *Who was this chipmunk, and who had they left behind? Had taking the chipmunk's body away from the site of their death made it impossible for their family to understand what had happened—to mourn?*

I was reminded of this chipmunk a decade later as I crossed the border into Vermont heading north from my temporary home in Connecticut to visit friends. After blurring past countless dead animals on the road, I was jarred back to attention by a large bloody rib cage on the side of the road. This rib cage must have belonged to a very large deer, or maybe even a moose—it was huge. The animal's full torso had been carved away, leaving the long curves of the ribs exposed, the white of the bone stark against the bright crimson blood and bits of flesh hanging from the ribs. Most of this animal was gone—and gone all at once by the looks of it. This creature's body had not been slowly eaten by animal scavengers—the blood was fresh and bright. Their body had been sliced up by a human, perhaps the person who hit them, or by another passerby. *We didn't want the chipmunk to go to waste.* I suddenly recalled my brother-in-law's words at that moment. *Go to waste.*

Eating becomes a way to not *waste* what (or who) could become a source of food for humans. In trying to avoid an animal killed on a road from *going to waste*, they acknowledge (at least in part) the unfortunate fact of the animal's death and the desire to do something to respond. The response, though, illuminates how these animals are situated in human imaginaries of where certain species, and their deaths, fit in a cosmology of life and death. Killed, even accidentally, by humans, these animals must be *of use*. Transformed in the instant of the collision from a free-living being to a dead potential site of waste, the animal must become useful either as a way to manage the feeling of helplessness that the driver might feel or as a way to avoid a more serious experience of mourning.

My brother-in-law and his girlfriend didn't know what to do with their grief over killing the chipmunk. There was not a prescribed way for them to deal with this grief that suddenly and unexpectedly confronted them. And so, they reverted to a more familiar way of relating to animals not easily categorized in our emotional inner worlds. This chipmunk was not an animal—like their dog Ginger or their cat Sophie—whom it would feel natural to mourn. But this was also not a species or an individual routinely rendered completely ungrievable, like those bred and killed for food or biomedical research. So many animal species are used and used up by humans, and often it is their usefulness that helps to excuse the troubling ethical questions their use raises. And so, for these deaths and the emotional response they prompt to be categorized and compartmentalized, they move silently and quickly out of the realm of grievability. Linda Monahan writes, "By reclaiming road-killed animals as food . . . otherwise superfluous animal killing [is inserted] into the established framework of killing animals for food. Reframing roadside bodies as usable to humans makes road-killed animals a happy consequence of car culture rather than a problem to be solved."[33]

Historically, animals killed on roads have also been seen as waste. In 1924, Fred Jackson, a game commissioner, implored drivers: "Is your mission so urgent and your time so valuable that you cannot pause for a second so that a life, no matter how insignificant it might seem to you, may be saved? Count the mangled feathered carcasses along the road and compute the utter waste therein. Have mercy and slow down!"[34] This plea reflects a sense of caring for animals' lives—a call to pay attention to those who are so indiscriminately killed, and whose lives and deaths may be routinely unregistered by drivers. His words challenge, to a certain extent, the anthropocentrism embedded in humans' entitlement to space and to fast transportation. And yet, even within what, to me, feels like a moving call to action to prevent animal death, his mention of "waste" reveals a conflicting agenda. *Waste* here can be understood as a *life wasted*, in other words, a meaningless and avoidable death. But Jackson was also a game commissioner when he wrote this, and so, at least part of his job would have been to engage in conservation efforts to manage populations of "game" animals for hunting. In this role, his mention of "waste" takes on a layered meaning. The death of animals on roads becomes a waste in the sense that they would have been useful for some other purpose—for being alive and available to be hunted.

Thinking of animals killed on roads as going to waste can be understood in other ways, too. There are multiple meanings of *waste*. As a verb, in the way that my brother-in-law used it, *waste* means to *fail to make good use of*. Again, the focus here is on a missed opportunity for *use*—one of the primary ways we conceptualize the place of animals in human society. Recuperating the animal's death as *useful* can be a way of trying to make sense of the senselessness of that death.

Going to waste also acknowledges where animals killed on roads are *going*, the end point for animals collected on roads—landfills, composting systems, or rendering plants. These animals become *waste*—"unwanted or unusable material, substances, or by-products."[35] And, as waste, they are easier to avoid mourning as they are transformed into *material, substance, by-product*. Waste is excess, detritus, something to be managed and disappeared. And so, these animals are routinely disappeared: driven over again and again until all that remains is a dark stain on the pavement, collected and disposed of by workers contracted by the state or municipality, hauled away as food by an opportunistic passerby.

In Seattle's International District, we are stopped at a traffic light in the left-turn lane. Eric is driving, and I'm in the passenger seat. My gaze is wandering ahead of us to distractedly observe the two lanes of forward-moving traffic. Their light is green (ours is still red), and the cars continue, leaving the road ahead clear. As the last cars clear the intersection, I see a frenzied blur of gray tumbling across the street. The blur is a pigeon, badly injured—probably hit by one of the cars. The pigeon is struggling frantically to stand or fly, to move out of the road, to make sense of what has happened to their body. I move to open the door to run out into the intersection to help, and I hesitate for a moment, looking to Eric and then back to the street. In the instant my hand starts to pull the door handle, a car comes racing by on the right and smashes the bird flat. The car careens on along its path, not braking for even a moment; it's possible the driver didn't even notice they ran over the bird. The pigeon is left crushed in the road, the only movement a flutter of some stray feathers in the breeze.

My mind is a flurry of panicked and nauseated thoughts: *Could I have gotten there—prevented the car from killing the bird—if I'd moved more quickly? I shouldn't have paused. Why did I pause? Would the bird have been able to recover from those first injuries? Would a vet even have been willing to treat them? What kind of pain would they have experienced in those last moments? Who witnessed this life being violently and unceremoniously snuffed out? Were there other humans observing from inside their cars or houses?* Life on the surrounding street looked as though it was going on as normal. *Did the pigeon have a companion or companions who watched their injury and death? Pigeons mate for life. Where was their partner? Who did this pigeon leave behind?*

Animal deaths on roads overwhelmingly occur in rural places, and so it makes sense that much writing and attention to these mortalities focuses on highways and rural roadways. However, living in urban environments for most of my life, I have witnessed a significant number of animals killed on roads and streets in the heart of the city. As I was writing this chapter, I walked up the street from my home, heading to the public transit stop—not far, actually, from where I had witnessed that pigeon's death so many years before. I was crossing the street, and I felt a punch to the gut as I saw the crushed but still recognizable body of a pigeon—very recently killed—in the middle of the road, bright blood oozing from a mass of gray and iridescent feathers. A large flock of pigeons milled around nearby. I wondered what they might be thinking and feeling. Just as I had that thought, a couple of teenagers, who were walking past the flock, both kicked the pigeons closest to them. The flock scattered, moving farther away from their dead companion in the road.

For years, in teaching undergraduates about subjects related to human-animal relations, I often covered the topic of animals killed on roads. In one of my classes, I asked the students to imagine the following: *You're driving along, and you see a form ahead on the side of the road. You're aware immediately that it's a dead animal. The shape and form are recognizable to you as a dog. You hold your breath as you come closer. You feel sick. You imagine the dog's family wondering where they are—putting up signs, calling the local shelters, posting on social media—hoping against hope that their companion will come home. Maybe you relate intimately to this experience because you, yourself, have lived with and loved a dog and can imagine the panic and sorrow at realizing they're lost, and then the gut-wrenching grief that you would feel at learning that they had been hit by a car and killed. You're thinking all of this in a rush of feeling in that moment when you see the animal ahead. But then, as you drive closer, you realize that it's not a dog but a large raccoon. What do you feel?*

Immediately, without pause, one of the students blurted out: *Relief.*

There is a hierarchy of grievability related to animals killed on roads and the empathetic response they evoke: a dog or a cat perhaps more so than a coyote, a fox more than a raccoon, a squirrel more than a rat, a blue jay maybe more so than a pigeon. Species typically despised or associated with waste and detritus—pigeons, rats, possums, raccoons, or a whole range of insects—typically do not warrant a second thought when we see them dead. Perhaps seeing a rat smashed flat or a pigeon crushed in the thoroughfare is more likely to elicit feelings of disgust than grief. Our attention may rest for a moment on these lifeless bodies but is then quickly diverted to something else: our phones, a passing car, a loud conversation across the street. And we've forgotten the remains we've passed, forgotten our disgust or perhaps a pang of regret, and these remains get driven over again and again until cars, rain, and sun erase any physical remnants of this animal's life and death.

How might we prevent this erasure? Push back against the physical, emotional, and psychological tendency to disappear these animal deaths and their bodies? What work can mourning do?

Artists have offered one possibility. The photographer Emma Kisiel assembled a photographic project called *At Rest* that memorializes animals killed on roads. She arranged flowers, stones, and leaves around the bodies of animals in the places she found them, and then photographed them. Her work not only is intended to make grievable these individuals' deaths to human viewers but also aims to address "our human fear of confronting death and viewing the dead."[36] Confronting the dead and acknowledging our own and others' mortality are not easy things to do. So many avoid thinking about it, talking about it, or facing it as a reality, even when it is imminently upon us. Many people, in fact, die, still in denial that they are dying. Circles of family and friends can lose meaningful time at the end of life because they deny that their loved one is, in fact, dying. Confronting death, then, is something a lot of us are simply not very good at. And this manifests in a particular way when we are culpable in someone's death— even when framed as an accident. But facing death and acknowledging the profound loss that's occurred for us and for others—mourning these losses—can transform our relationships with others.

In one of the classes I taught at the University of Washington, after viewing and discussing Kisiel's *At Rest* during the previous class session, one of the students was very excited to announce that she had seen a memorial just like Kisiel's on the side of one of the small roads through campus. They came upon a tiny mouse surrounded by a carefully laid circle of flowers and leaves. One of the other students blurted out, "I saw that, too!" And a third student announced that they had been the one to create the memorial when they saw the dead mouse the day before. Who else had passed by this memorial, and what did it mean to them? Kisiel's art and the work it does in the world, then, can travel beyond those who encounter her photographs. It can become a living practice of honoring the dead, of making recognizable and grievable to humans the possibility of who and what was lost. Intentionally grieving or making grievable for humans those animals who might otherwise not be is a political act of acknowledgment of subjects of violence and disposability. Thus, a project that asks people to recognize what, or *who*, is grievable, and then to mourn them, is a powerful invitation and provocation.

L. A. Watson, a Kentucky artist and activist, offers a different kind of memorial. On a country road in rural Kentucky, a cluster of road signs are staggered along the shoulder, warning of wildlife crossing the road. These aren't like typical road signs. They're white silhouettes in the shape of animals commonly killed on Kentucky roads: a rabbit, a fox, a raccoon, a possum. These signs constituted Watson's *Roadside Memorial Project*. Watson asks, "If road signs are visible markers that reflect a human-centered economy, which values some but not all lives, how can both the memorial and the sign be reimagined as one and the same?"[37] Watson takes on a reimagining of road signs—simultaneously offering a memorial to those who have died and an active effort to reduce the number of future deaths. She points out that road signs warning of high-density wildlife crossings disproportionately alert drivers to the potential presence of larger animals, like deer, bears, or moose—species who pose a greater threat to human life and property damage than do smaller animals. These road signs, then, are meant to protect humans, ignoring the danger and risk to animals' lives. They are positioned high up on a post, closer to a human's eye while driving. Watson's signs, by contrast, are positioned low to the ground—still highly visible with their white reflectiveness, but they draw the driver's eye down closer to the height of actual animals.

When I followed up with Watson six years after she installed *The Roadside Memorial Project*, I asked what had happened with the project. She said that, over the years, she had moved the signs around "to keep people guessing as to what they would see or expect."[38] Some people asked Watson about the signs, curious about their meaning and purpose. Some of the signs were stolen or removed. But a consistent effect was that drivers along that stretch of road reduced their speed (the primary method of avoiding collisions). This kind of project, then, prompts drivers to *do something*, even if it's an inadvertent act like slowing down when you see something unusual. Watson's and Kisiel's artistic works ask viewers to acknowledge and mourn but also to do something tangible, however subtly—to memorialize, to slow down, and to make efforts to change their own practices.

Both of these distinct artistic forms of drawing attention to and honoring the dead do transformative work in the everyday world, too, through actively reconceptualizing for humans the reality of loss and focusing on singular casualties of human travel on roads. These casualties are overwhelmingly numerous, though, and sitting with this reality is uncomfortable, painful, and exhausting, especially without widespread ways to undo or prevent this harm and damage. To sit with this reality, and to hold it in the front of the mind and in the heart, is to be haunted—maybe perpetually—by these many losses.

I felt keenly this sense of being haunted for years following an encounter on another stretch of our cross-country road trip in 2006. I was driving, and Eric was in the passenger seat. We were barreling along a hot, dry Arizona highway. The thermometer read 115 degrees Fahrenheit. The air-conditioning was coming on and going off in fits and starts, providing moments of marginally cool air as respite from the sweltering heat. We rolled down the windows, hoping for some relief, and the blazing, dry wind whipped through the truck. Which was better, we weren't sure. The landscape looked at once deadly and breathtakingly beautiful. Desert stretched out on either side, sparsely covered in scrub and punctuated by iconic saguaro cacti.

As I was driving, I noticed what looked like dried leaves blowing across the highway as far ahead as we could see, where they turned into a mirage on the horizon of the road. Maybe it was because I was delirious from the heat, but it didn't register right away that there was no vegetation around us that would produce large, dried leaves. All at once, Eric and I realized, with horror, that the "leaves" were tarantulas, moving quickly across the road in droves, like a dark wave. We had been crushing countless tarantulas as we drove. I cringed, sick to my stomach, and kept driving, unsure of what else to do. I learned later that tarantulas migrate south, in large colonies, from Colorado and Nevada to Arizona and other parts of Southwest, in September and October each year. We were traveling in September and, without realizing it until it was too late, came upon this mass migration and killed an unknown number of these arachnids who were only trying to move south, their path bisected by a road—a seemingly benign (but profoundly deadly) feature of the landscape. And we were not the only car on the road, so the death toll would have been much higher than the massacre delivered by us alone.

These tarantulas haunt me. I can see them clearly in my mind even now—the way they moved, in a fluid, almost coordinated way, rhythmically, across the burning asphalt. We should have stopped, waited for the migration to finish, however long that took, and refused to crush any more of these creatures who were just living their lives, moving through the world in a way they did every year. But we didn't stop. Our destination, schedule, and discomfort in the heat caused us to keep going, moving on to cooler climates. In the years since, though, I've thought often about this experience and how our own culpability was easily eclipsed by priorities that didn't involve an attentiveness to the experiences and lifeways of those tarantulas. Practicing attentiveness in that moment would have meant stopping the car and waiting until they passed, appreciating the beauty of their delicate bodies moving across the road in waves.

My earliest memory of culpability and grief in killing an animal with a car is from when I was seven or eight. My mom and I were driving down a highway and suddenly a monarch butterfly appeared, crumpled, on our windshield, lodged into the top of the windshield wiper. We both let out a gasp, and I begged her to pull over. I was sure we could save the butterfly. She pulled off on the side of the road, and we got out of the car. I waited while she extracted the broken monarch, ever so gently, from the windshield wiper. She handed me the creature, who was still alive. I cupped my hands and watched as the orange and black wings rotated, the butterfly trying to right themselves and stand up. But their whole body was askew, contorted in ways that looked irreparable, even to me, a child who knew nothing about butterfly physiology. Soon the movement stopped, and the monarch's wings settled, finally, against my palm. I choked out a single sob and walked away from the car, over to the grass, and chose a spot at the base of a plant with clusters of pink flowers—milkweed, my mom said it was—the favorite food of monarch caterpillars.

I laid the beautiful, broken body gently on the ground, and we stood there for a quiet moment before heading back to the car. As we approached the car, I noticed that the front grille was covered with layers of dead insects we had killed while driving. I remember my eyes welling up with tears—it was too much. Too much death. Too much killing for which we were responsible. Too many insects to pluck from the car grille, lay to rest, and mourn. How are you supposed to hold that much grief at once, that much culpability for that much violence, all for a car trip? I don't even remember where we were going.

For animals who migrate, like tarantulas and monarchs, roads can be particularly devastating. The phenomenon of large numbers of a species migrating across a roadway all at once results in a much greater number of deaths than occur among animals who are moving around their habitat as individuals. The state or federal entities that manage most roadways do not close them or change their use and availability during migration periods. However, one well-known example of accommodating annual migration patterns exists in southern Illinois, in the Shawnee National Forest. The Shawnee National Forest is a highly biodiverse ecosystem that has been preserved as a place for plant and animal species to flourish undisturbed by human development and activity, and it's also a site designated for researchers to study these species in situ.[39] There is a 2.5-mile stretch of road in the park across which large numbers of snake and amphibian species (some of whom are endangered) migrate in the spring and fall each year. Mortalities on this road were significant after it opened to traffic in 1972. Noticing these mass deaths from traffic collisions, park officials decided to close the road for two months in the spring and two months in the fall to dramatically reduce these mortalities.[40] Now, the migration itself is a draw for tourists and researchers, who can access this stretch of road only by foot during migration periods.

Perhaps this is an example of humans intentionally changing their patterns of movement to accommodate animal movement that might bring some hope. However, this effort works in part because the site of this migration is a national forest, a place where attention to wildlife and protection of the ecosystem is implicit. Closing this section of road to traffic does not dramatically impact major flows of traffic traveling for commerce, commuting, and other purposes on major roadways. While the road closure may be an inconvenience for those using or moving through the Shawnee National Forest, it is not a major disruption for high-volume traffic. Would this be at all possible or probable on a major interstate, for instance?

There are, in fact, highway projects completed and underway that aim to respond to the problem of motor vehicle collisions with animals and ultimately reduce the number of animals killed on roads. Mitigation of animal mortalities on roadways is not at all a new concern. Foresters and ecologists in the early twentieth century were consistently worried about the issue of animals killed on roads, especially as the speed of vehicles and the number of vehicles on the roads both increased. The 1960s and 1970s saw escalating concern as deer and other large animals posed a significant threat to human life and property via motor vehicle collisions. Researchers began identifying hot spots of deer activity and focused their energies on exploring ways to mitigate the effects around those zones.[41] Automobile collisions with wildlife have long been framed in economic, cost-benefit analysis terms: *What were the costs of mitigation versus the mounting costs of collisions?* This became the site of ongoing research and began to shape policies around new road construction and adaptations to existing roads, and is still a central argument today for developing new and more widely adopted mitigation strategies.

Driving east on I-90 from Seattle, there's a fifteen-mile stretch of highway in the Cascade Mountains that has been the site of a road redesign to accommodate wildlife crossings. Animals travel north–south in this corridor; humans travel east–west on the interstate. As a result, this section of road has been the cause of exceptionally high wildlife mortalities, harming not only animals who are injured or killed but also humans and their property (cars). The interstate bisects a major wildlife corridor for the region—one that, over the years, has been intentionally cultivated to increase habitat connectivity. This region of wilderness was fragmented when private landowners bought up parcels of land. Once the importance of a comprehensive wildlife corridor in the state was understood, the US Congress, individuals, and Washington state worked to purchase this land back, restoring one hundred thousand acres along this corridor.[42] The interstate, though, posed a real problem. Restoring connectivity through land purchasing was an important step, but the survival of many species was still at stake where the interstate cut this corridor in two.

In an unprecedented collaboration among many stakeholders, including the Washington Department of Transportation, wildlife biologists, foresters, and others, the I-90 Wildlife Bridges Coalition came together in 2004 to catalyze and formulate a comprehensive plan to construct wildlife underpasses and overpasses to support safe passage across the highway. These wildlife crossings were wrapped into more conventional road construction projects—road widening, repair, and general improvements to human travel on the interstate. Wildlife underpasses create passages under the roadway to allow land and aquatic animals to pass beneath the road safely. Some are narrow, tunnellike crossings, whereas others are wider and more open, with the road forming a bridge overhead. Overpasses, by contrast, are land bridges constructed over the highway that are wide, gently sloped, and planted with native flora to encourage animals to feel comfortable using them. The aim is for these bridges to feel to the animal like they are walking through the wilderness and not like they are crossing a human-made bridge over a roadway. The project involved the construction of several underpasses and a bridge overpass, as well as culverts to allow fish and aquatic life to move easily through waterways. Initial plans for the project included a more expanded network of wildlife passages, slated to be completed by 2029, and with a budget of $1 billion, but in 2018, the I-90 Wildlife Bridges Coalition shut down, leaving its completion up in the air.[43]

Underpasses and overpasses have been successfully constructed and put to use in other places around the world. One of the most prominent examples is in Canada's Banff National Park, where planning and construction has been ongoing since 1994 to create underpasses, overpasses, and fencing to try to prevent collisions with animals on the Trans-Canada Highway. In addition to a robust series of overpasses and underpasses, fencing has also been constructed along the highways to prevent large animals from trying to cross in between the formal crossings. Although these fences dramatically alter animal movement and reduce a smooth connective flow of habitat, they also help to prevent mortalities. Fencing is, importantly, used in conjunction with overpasses and underpasses, working as a funnel to encourage animals to move toward and through or over the highway crossings constructed for them. These efforts have seen success where they have been implemented, causing a marked reduction in motor vehicle collisions with wildlife.

As is the case with the I-90 plans, these projects are often packaged with more conventional highway improvement plans—for instance, they are wrapped into major roadway repairs or, frequently, become part of highway widening projects meant to accommodate higher volumes of traffic. Highway-widening projects not only cut farther into often-critical habitat but also increase the likelihood of collisions with wildlife by reducing the chance that an animal will make it safely across so many lanes of traffic. Jackie Corday from Missoula, Montana, shared her story with an organization called Animal Road Crossing:

About 10 years ago the main highway that I commuted on to work expanded from 2 lanes to 4–5 lanes. Soon thereafter, I watched 2 collisions with deer happen—in each case the deer made it across 3 lanes and then got smashed in the 4th lane. The two extra lanes significantly decreased the deer's chances of crossing safely. It was very heartbreaking to witness those deaths and very stressful knowing that I could go slower and pay attention in order to avoid hitting wildlife, but the vast majority of drivers do not.[44]

The primary objective of transportation improvement projects is to enhance flow and ease of vehicular movement across space, and to increase the safety of drivers as they travel. Wildlife crossings are couched in terms of improving human safety while driving; fewer collisions with animals on roads reduces risk to drivers. Cost-benefit analyses show that it is usually economically advantageous to invest funds into these collision-reduction projects to decrease the substantial cost of collisions and all the expenses that follow from them. Animals themselves tend to be peripheral to these concerns. The dominant anthropocentric worldview surrounding roadways, transport, and safety predominantly frames human life as precious, to be protected at all costs, and animal life as disposable, except where it intersects with priorities around human safety and well-being and economic considerations.

Cost-benefit analyses are also overwhelmingly focused on large animals (deer, elk, moose) who do significant damage to vehicles and sometimes cause human deaths when these collisions occur. Societal concern is rarely expressed over smaller animals killed by cars, unless these animals happen to be part of an endangered species. Some species already threatened or endangered are drastically impacted *as populations* by motor vehicle collisions. For instance, a 2008 study by the Federal Highway Administration identified twenty-one threatened and endangered species who are particularly vulnerable to population decline because of motor vehicle collisions. Threatened or endangered mammals that they studied include the Key deer, the San Joaquin kit fox, and the Florida panther; reptiles include the American crocodile, the desert tortoise, and the eastern indigo snake; amphibians include the California tiger salamander, the flatwoods salamander, and the Houston toad; and birds include the Hawaiian goose, the Florida scrub jay, and Audubon's crested caracara. While other factors contribute to the rapid decline of these species, such as agricultural development, urbanization, and suburban sprawl, motor vehicle mortalities are a key deadly impact for these vulnerable species as well as for species not labeled *threatened* or *endangered*.

Whether or not a population is threatened by extinction (and this threat varies based on species and surrounding environmental factors), motor vehicle collisions with wildlife catastrophically affect the *individuals* of any species they kill, and this death reverberates through the animal community surrounding this individual. Humans have a persistent tendency to scale out to the level of the population when talking about animals—especially animals with whom they do not have intimate relationships. It is this abstraction from the individual, though, that renders them less grievable by humans and allows for bypassing ethical concern for effects on single animals. Acknowledging the individual animal killed on the road—or the individual who *might* be killed on a road—can prompt us to feel and attend to grief and our own culpability. We might become haunted by these deaths and losses. These emotional responses and hauntings emerge from witnessing this loss in an intentional way, and they can engender a rejection of our tendencies to ignore, look away, or forget the corpses we pass as we zoom by, en route to somewhere else.

Roadway mitigation projects, practices to prompt mourning and honoring animal deaths on roads, and individual projects to inform, move, and inspire drivers to take greater care as they move through space all draw important attention to the plight of animal deaths by automobiles. These projects manifest in a context where roads, automobiles, and high-speed travel are taken as a given, and this is understandable, given how entrenched these norms are in a consumer culture located within structures of capitalism and settler colonialism. But I wonder, what might it look like to engage in a more radical reformulation of how road travel represents a sense of entitlement to practices that improve the ease and convenience of *our* lives over the survival of others? How might we work to undo the damage wrought by roads and automobiles, and the logics that underwrite this damage?

Meaningful repair and transformation in these contexts "will require more than isolated wildlife corridors and overpasses and other small gestures toward the development of a more animal-friendly auto infrastructure. Indeed, it will require a wholesale political challenge to automobile dependency, the auto-industrial complex, and—more broadly—the socially, psychically and environmentally corrosive logic of commodification itself."[45] Capitalism thrives on efficiency and the speed of the production, circulation, and consumption of commodities. The proliferation of cars, enabled by Ford's assembly line, is an iconic example of this guiding logic. Perpetual growth, too, is central to how success is framed in a capitalist economy. But there is now a mounting understanding, globally, of the devastating effects of this unfettered growth and accumulation on the environment, on people's lives and deaths, on animal individuals and populations. Within this context, how could the phenomenon of animals killed on roads be ameliorated, or even eliminated?

Slowing the speed of travel is one effective way to reduce mortality on roadways. Reducing speed, however, goes against the high speed and efficiency needed to accelerate capital accumulation. Limiting travel altogether during peak times of day when collisions with animals on roads tend to occur (often dusk and dawn) could be another way to reduce mortalities. Reducing speed limits on roads or limiting travel during certain times of day might mean a decrease in overall circulation of goods and commodities that move over and across landscapes in motor vehicles. It might mean an overall slowdown of consumer culture. It might mean a reconfiguration of the typical workday to accommodate different, more multispecies, interests. It might mean a not-insignificant inconvenience to travelers. And it might mean that we observe more carefully those landscapes, those animals, and those human communities that we pass by as we travel. Perhaps a reduction in speed might lead to less abstraction and less ignoring or forgetting.

These ideas fit within a growing global movement for degrowth, an effort dedicated to intentionally limiting economic growth, shrinking markets, and indeed, moving away entirely from a capitalist economy. At an international conference held in Paris in 2008, *degrowth* was collectively defined as a "'voluntary transition towards a just, participatory, and ecologically sustainable society,' making clear that a downsizing process was necessary [to build] a society organized around sharing, simplicity, and solidarity, rather than the profit, efficiency, and competition inherent to capitalism."[46] Roads, travel, and the harm they do could be one site of potential transformation oriented within a degrowth model—one that attends to a multispecies practice of global well-being.

These suggestions, however—to reduce speed or limit travel during certain times of day—even as they might push back on the guiding logics of capitalist economies, do not address more fundamental questions about the existence and maintenance of roads in the first place. And these questions are, perhaps, much more difficult to address, and likely much more uncomfortable for those of us who are settlers on stolen land. A true transformation of the violence of multispecies relationships as they relate to roads, automobility, and travel would necessitate making this issue a part of the active work of decolonization—a radical undoing of the bedrock of colonial society and the violence that flows from it, and an opening for more caring and thoughtful relationships with the land and with other humans and animals. Framing decolonization here as an act of doing something tangible in the world necessitates a repatriation of stolen land.[47] A commitment to the decolonization of roads and landscapes of travel and movement would align with the LANDBACK movement, initiated by the NDN Collective, to "put Indigenous lands back into Indigenous hands," where the return of ancestral lands enables a powerful move for Native sovereignty. Undoing or repairing the damage done by roads would be one part of a broader land repatriation project, like the one called for by LANDBACK not only to the First Peoples around the globe but also to the nonhuman animals and ecological systems from whom they were seized in the first place and to whom their seizure continually does damage.

One example of what greater Native sovereignty might look like in the context of roads specifically could be highlighted through the case of a project oriented around the repair and management of Highway 93 in western Montana—a two-lane highway in severe disrepair through the 1970s. In the 1980s the Montana Department of Transportation (MDOT) embarked on a plan to transform the narrow road into a four-lane highway that would cut even more aggressively through the landscape. The highway was notoriously dangerous and narrow, a site of frequent human-animal collision, injury, and death—and it was almost entirely located on the Flathead Indian Reservation, the traditional lands of the Confederated Salish and Kootenai tribes. As such, when plans were announced by MDOT to construct a much larger highway in its place, the Tribal Council of the Salish and Kootenai (TCSK) objected, citing the profound harm this smaller highway had done to their lands and environment and the intensification of this harm that would be wrought with the construction of a larger highway.

The Federal Highway Administration, MDOT, and TCSK came to an impasse over how to proceed that held up the project through the 1990s. Then, in 2000, they reached a compromise that suited the various stakeholders. A landscape architecture consulting firm brought in to formulate a design for the project took all considerations into account and proposed the concept of the Spirit of Place for the project, As the architects explain:

The design of the reconstructed highway is premised on the idea that the road is a visitor and that it should respond to and be respectful of the land and the Spirit of Place. Understanding the Spirit of Place—the whole continuum of what is seen, touched, felt, and traveled through—provides inspiration and guidance, and leads to design solutions uniquely suited to the special qualities of the place. The Spirit of Place includes more than just the road and adjacent areas; it consists of the surrounding mountains, plains, hills, forest, valley, and sky, and the paths of waters, glaciers, winds, plants, animals, and native peoples. The Spirit of Place encompasses the entire Mission Valley, Mission and Salish Mountains, Jocko Valley, and Rattlesnake Divide. This broader environmental continuum has distinct landscapes like large outdoor rooms, which the existing road bisects.[48]

At the heart of the Spirit of Place was the cultural heritage of the Salish and Kootenai tribes. One specific way that their cultural heritage would be honored was in and through the consideration and protection of native wildlife who were central to the cultural traditions and beliefs of the tribes. Like the I-90 wildlife corridor project in Washington and the one in Banff National Park, Highway 93's redesign incorporated wildlife crossings to protect many at-risk species and human travelers on the road, in this context orienting itself around a consideration of Indigenous sovereignty as the new road was planned and constructed.

Such a collaboration between Native and non-Native stakeholders holds some promise for potentially decolonial ways of designing and redesigning roadways into the future, despite the complexities surrounding this federal, state, and tribal collaboration. LANDBACK's call for "putting Indigenous lands back into Indigenous hands" would at the very least involve centering Native sovereignty in formulating how, where, and with what effects roads are constructed and maintained. This could be the beginning of what decolonization might look like in relation to roads, their histories, and their futures.

And yet, to push further, I'm interested also in what it would look like for settler communities and institutions to cede power and sovereignty not only to Native communities but also to other species. I don't say this in order to sidestep the active work required of settler subjects and institutions to cede sovereignty to Native human communities. Instead, I pose these questions to prompt some conversation about what it could mean to decenter humans more generally: What might it look like to follow other animals' lead in their own self-determination about the places they live, the spaces they share and cocreate, and the ways they flourish in their multispecies communities? What might it look like for humans as a species to observe, attend to, and honor what animals might be showing us on a more widespread scale about what they need and how they want to live, as some Indigenous and non-Indigenous humans are already doing? Might this not be one part of the work of decolonization—allowing animals to define what a decolonial future might look like?

This work may seem impossible to imagine, especially for those of us who enjoy so thoroughly and unthinkingly the conveniences and privileges that are so normalized as to feel natural, both as settler subjects and as humans in general. It would involve a dramatic surrender by settler, capitalist society of convenience, of wealth accumulation, of entitlement to space, of human and nonhuman others as sites of extraction and commodification, and of human supremacy at its very core. As with the other multispecies sites of everyday haunting and witnessing explored in this book, we are left with uncomfortable questions and even more uncomfortable answers. What kinds of journeys must end here?

3

The Scent of the Spectral

To orient ourselves around scent is a haunting, a whiff of memory, of history, of the violence we've caused and experienced. It's something ephemeral, cleared in a moment with a breeze but not forgotten.

Long before I could hear or see the animals, I could smell them. I entered the University of Washington's massive biomedical research complex through a back loading dock, following a staff person involved with animal care at the facility (I'll call her Allison) through a maze of nondescript hallways. The corridors on this ground level of the building had exposed ductwork and pipes running along the ceilings, and the yellow-hued fluorescent lighting flickered on and off at irregular intervals. Lining one side of the hallway were tall carts stacked with plastic rodent housing boxes. Save for the soiled wood shavings lining the bottom of the boxes, they were all empty. Several carts held larger metal cages, also empty. The smell of soiled bedding absent the animals had a ghostly quality. *How many generations of mice and rats had lived in those small plastic boxes, suspended in a perpetual state of waiting—for what? Death? A livable life that would never come? Who had been housed in those larger cages? Whose absences did those empty cages contain?*

I felt a prickling sensation on the back of my neck, the eeriness of this place and its absences sending a shiver down my spine. I shook off the feeling and joked to Allison that, this being my first visit to the animal housing unit, I'd never find my way back out of the building. She assured me that she would walk me back out and explained that for security reasons I must be accompanied at all times during my visits to this facility.

We rode up to the sixth-floor vivarium in a creaking service elevator, and I waited while Allison scanned her key card to admit us into a long white hallway. Visually unremarkable, it looked like any interior institutional corridor: white walls, beige linoleum floors, and on either side, rows of solid windowless doors with stainless steel kickplates. Each room had a laminated sign on the door, identifying the species inside with text and a picture: *Dogs. Rats. Mice. Rabbits. Ferrets. Amphibians.* I thought, looking down the nondescript hallway in either direction, that we could be anywhere—there were no windows, no geographically identifiable features once inside. This could be any animal research facility in any number of countries where nonhuman animals are widely used in biomedical research. Or without the animal-identifying signs on each door, it could be a hospital, a psychiatric facility, a nursing home, or another space dedicated to the institutionalized management of life.

Here, in this hallway, the absences that haunted the ground floor hallway turned into presences rendered real in the smells and sounds of the sixth floor. If the odor of animals in confinement was noticeable on the ground floor, it was magnified tenfold on the sixth floor. The sharp, stringent smell of disinfectant, too, burned the inside of my nose as it tried unsuccessfully to eliminate or mask the other odors. Dogs barked and howled from behind closed doors.

I was there to meet a beagle whom Eric and I had agreed to adopt. I followed Allison down to the far end of the hall, where she unlocked the door to an empty exercise room situated between two dog-housing rooms. She asked me to wait there and then disappeared through the door on the left, coming back a few minutes later dragging a small, unwilling beagle on a leash behind her. Allison introduced me to Donut, as she was called by the animal care staff, for the white on her muzzle that looked like she'd gotten into a box of powdered sugar donuts. By the researchers, and in her records, she was referred to as DAE-1, identified by the letter-number combination tattooed in rough blue ink on the inside of one velvety ear.

Allison left us alone to get to know each other. The frightened beagle cowered in the far corner of the room. Her small body shook uncontrollably, her long ears quivered, and her bloodshot eyes followed my every move. I sat quietly and waited. Dogs barked, howled, and whined from the rooms on either side of us. I spoke softly to her. She shivered and shook. I sat quietly and waited.

Suddenly her nose started sniffing ever so slightly, her head tipped up, as she smelled the air more vigorously. Finally, she gingerly tiptoed over to me and smelled my outstretched hand. I was sitting on the floor, cross-legged, and she sniffed at the knee of my jeans, my hands, my shirt, likely reading the smells from the places I'd been that day, the myriad scents at home, my anxious sweat, the dog and two cats Eric and I lived with, a bit of crusted oatmeal on my shirt from breakfast, perhaps even my intense longing for her to be at ease with me. I wondered what she was smelling and what knowledge was coming to her through that powerful nose. I gently stroked the top of her head, softly scratching behind her ears. She accepted my touch, although she watched my every move warily.

I sat with her in the room until Allison came back to retrieve me, and I waited while she took Donut back into the kennel where she was housed. When Allison opened the door to the room, the barks and howls I'd been hearing were shockingly loud as they reverberated in the mostly empty room. I followed her to the door and saw that it contained two adjacent four-by-eight-foot kennels made out of chain-link fencing from floor to ceiling. Donut was returned to the kennel on the right side, joining a pack of four other beagles. The kennel on the left contained three more. All of them looked alike, a fact that I would later learn was a result of the way they are bred specifically for research by one of a few large breeding facilities in the United States.

As we left the beagles behind, I asked if there were dogs housed in the room opposite the beagle room. Allison opened the door to that room to show me the larger hounds confined in individual cages. Allison explained the increasing difficulty of finding adoptive homes for the many dogs, ferrets, and occasional cat the lab uses for human biomedical research. Except for the odd vet tech, student, or researcher who might adopt a rat, mouse, rabbit, amphibian, fish, pig, or sheep, attempts were not made to adopt out these other species. Even as it was, the adoptions that did occur, of dogs mostly, had been initiated and facilitated entirely by Allison, a passion project of hers that began and ended with her employment at the research facility. They also had to be done entirely by word of mouth, and with promises of complete discretion, to ensure the fact that the university was testing on dogs would not be publicized.

As soon as we got the go-ahead to bring Donut home, after she had been spayed and healed from the surgery, Eric, my sister, and I went to pick her up. Allison and another vet tech met us outside the loading dock with a large crate covered in a sheet. We lifted the crate, containing one terrified beagle, into the back of our station wagon and left the lab behind. In the car, she shivered and shook in her crate, vomiting before we were halfway home. We spoke softly to her, reassuring her that she was going to be all right, telling her she was going home to a new life. Leaving her lab name Donut behind, we named her Saoirse, an Irish Gaelic name meaning "freedom."

Before meeting Saoirse, I knew very little about the experiences of animals bred and used for biomedical research. But getting to know her led me to consider questions about human entitlement to the bodies and lives of other species in medicine. What does it mean to weigh human life against animal life again and again in medicine and always settle on the side of prioritizing the potential benefits to humans? How do the high stakes of medical advancement obscure the complete appropriation of other animals who do not consent to their lives being consumed for the perception of human benefit? How much does animal research actually translate into human medical advancement? What if we rescinded our sense of human entitlement to other species and focused instead on our shared multispecies flourishing, well-being, and care? How can individuals like Saoirse help us to ask and answer these questions?

Animals have been widely used in medical research for over two millennia; some of the earliest recorded instances of live animal experimentation (vivisection) date back to research performed by Aristotle in the fourth century BC, Erasistratus in the third century BC, and Galen in the first century AD. Throughout this long history, animals have been used as proxies for human subjects in the early stages of medical and pharmaceutical developments. Animals have been bred, lived, and died in experiments for the development of vaccines, surgical and organ transplant techniques, treatments for chronic diseases, and a range of other medications and medical procedures. Generally, once techniques, treatments, or medicines have been proved to have some success in experiments on animals, they can go on to be tested on humans.

Human medical testing itself has a long, fraught, and racist global history. From the Tuskegee Syphilis Study to the use of enslaved women in the development of gynecological techniques to the ongoing use of prisoners for biomedical research, widespread use of racialized and disenfranchised human populations in harmful and sometimes deadly studies is a pattern that stretches across the past and present of medical and scientific progress.[1] Committees at the level of the research institution, like the institutional review board (IRB) for research on human subjects, are meant to oversee research design and protocols, ensuring that research is conducted in accordance with ethical guidelines. At the core of these reviews is the issue of informed consent—that in most cases humans are required to be fully informed about the potential effects of the study and give their consent only after this information has been provided. However, there remain serious problems with whether and how consent is adequately obtained in all cases with humans.

Questions of consent, vulnerability, power, and the inadequacy of meaningfully protective legal structures create a troubling landscape in medical research. Laws are problematic systems in this regard—the passage and existence of laws meant to protect people and animals in inherently compromising situations, like medical experimentation, presume that the harmful practices of experimentation on living subjects are, at their core, acceptable and, applied with the correct restrictions, can be ethical. This is not to say, of course, there shouldn't be laws protecting vulnerable subjects. But laws—perhaps especially *protective* laws—create a status quo wherein a practice has been reviewed, considered, and determined to be acceptable so long as it's done within particular legal guidelines on welfare, safety, and ethical conduct. Rarely do these kinds of laws result in a complete prohibition of the use or exploitation of vulnerable populations, instead regulating that use or exploitation through a rebranding—in both discourses and material practices—of something unethical into something that becomes widely accepted. This has been true in the context of human medical experimentation, and it is the case in animal experimentation.

In the context of laws addressing and setting standards for the welfare of animals in biomedical research, the Animal Welfare Act was adopted in 1966.[2] The law was meant to address these issues and included in its coverage protection for dogs, cats, nonhuman primates, guinea pigs, hamsters, and rabbits. The act *excludes* all mice, rats, birds, fish, farmed animals, reptiles, amphibians, and cephalopods, who do not count as animals in the letter of the law.[3] These species, however, constitute the overwhelming majority of animals used in research. Saoirse is just one of the forty-seven thousand dogs used each year in US animal research. And those dogs are just a fraction of the animals whose lives are reported to be used in research each year.[4] The numbers of animals reported represent only those covered by the Animal Welfare Act. As a result, the figures of animals used in research that are captured in the numbers reported to the US Department of Agriculture (USDA), the government body that tracks these statistics, are misleadingly low.

A broader accounting of the number of animals used in research is difficult to ascertain. Because only animals who are protected by the Animal Welfare Act (5 percent of animals in research) are documented, the 95 percent of animals who are excluded from the act—primarily rats, mice, and other rodents—go unreported in animal use records and unprotected by federal animal welfare law. Officially, in 2023, the USDA reported that roughly 744,065 animals were used in biomedical research.[5] However, if all of the typically excluded species were included, conservative estimates cite roughly 14.5 million used annually, but a 2021 study estimated that rats and mice alone accounted for 111.5 million.[6] Globally, it's estimated that 192 million animals are used each year in scientific purposes.[7] As is reflected in these widely variable estimates, without reporting on all animals used, it is impossible to confirm exact numbers.

Absent from these counts, their lives are all but erased. Numbers count the presence and existence of someone who lives and dies for research, and so accurate and transparent counting is important. But numbers also have the potential to be abstracting. It's easy to forget that these are individual, singular lives—that all of these animals were *someones* with complex emotional inner worlds and the desire for social connection, care, and lives not oriented around being held captive, experimented on, and killed in the name of research. When their numbers are not recorded or reported at all, they become all the more abstracted, their lives not even reduced to numbers on a page. To be abstracted is to be "absent in mind."[8] These 14.5 million animals in the United States alone, and their very real presences in the lab, become absent from the minds of most of those beyond the lab. I'm reminded of the empty cages in the animal research facility—the absences of those rats, mice, and other animals who lived and died, were used and held, were present for themselves and each other, until they were studied, killed, and discarded, their bodies incinerated as medical waste, leaving behind only the scent of their soiled bedding as ghostly absences, until this too is discarded, replaced with fresh bedding and new animals who will soon become ghosts themselves.

Saoirse arrived at our home terrified. She would sit like a rag doll anywhere we put her in the house—still, unmoving, except the constant trembling and shaking that was a persistent behavior in those early days. When there was a nearby hiding place available, like under a table, she would tuck herself as far back into a corner as she could. Saoirse's hypervigilance exhausted her, and the exhaustion amplified her anxiety. These can all be signs of trauma in dogs; as humans do, dogs can suffer from traumatic experiences and carry post-traumatic stress disorder (PTSD) far into their future.[9] Dogs are not the only species apart from humans who suffer from PTSD and other trauma responses; many species, including rats and mice, experience lasting responses to trauma they have undergone in lab research and other contexts, and these trauma responses can increase and intensify over time.[10] Like humans, nonhuman animals suffering from PTSD and other forms of trauma can be triggered by certain events, sounds, smells, or encounters long after a traumatic experience has occurred.

Immediately after Saoirse came to live with us, it seemed like the whole world and everything in it was a trigger—so overwhelmed and frightened was she. Everything was new for her, and even those things that ultimately ended up being positive experiences were too much. Saoirse's fear was a full-body experience, taking over every aspect of her encounter with the world. Her fear itself, in fact, has an odor to it. Over the years of our lives lived closely together, I've come to know the perceptible change in the smell of Saoirse's breath when she is afraid as she huddles against me for comfort, her body hot and tense. Even now, years after her release from the lab, her fear is viscerally made manifest in these and other ways. Our task has been to help her find the things that make life tolerable, places and moments where and when she can be at ease. She discovered, for instance, that she felt safest on the couch, burrowed in a pile of blankets. When she got frightened by something, she would dash to the couch, tunnel into the blankets and pillows we kept there for her, and within minutes would feel safe enough to fall asleep, snoring loudly.

When she left the lab, she had never been outside. She'd never seen grass or the sky. She'd never been on a couch or a bed or in a house. She'd never met a cat. She'd never seen a squirrel scamper across the sidewalk. She'd never tipped her head up in the fresh summer breeze to smell whatever the wind carried to her nose. The first time we took her outside, I carried her out to the front yard. She was apprehensive about the feel of the grass under her paws, looking down often and picking up a paw to see what she was standing on. When we ventured out of the yard, it took us thirty minutes to walk around our small city block. Her tail was tucked tightly against her belly in a canine signal of fear and submission, and she smelled nearly every inch of the parking strip the whole way. When we got home from the walk, she didn't know how to get up the three stairs to the house because she'd never seen steps living in the lab. We taught her by lifting her front paws up onto the first step and then her back paws, repeating this until finally she learned to do it on her own. Every day was a negotiation, an experience of learning and adapting to this new world she lived in.

Above anything else, the thing that helped Saoirse transform her trauma was encountering the world through *scent*. The smells of the outdoor world around her animated her, opened up her reality, brought her out of her fear, and ultimately helped her to heal. She would catch a scent, and we could see her forget for a moment to be afraid. She would follow the scent, her whole body-mind coming to life with the deep inhalations that allowed her to take in the rich complexity of what she was smelling—messages to her about life unfolding all around her, messages that transformed those fleeting moments of fear forgotten into longer and longer expanses of time.

It was astonishing, in fact, to witness the power of scent as it transformed Saoirse from her fearful, cowering, shivering self when she first came home to an intrepid, adventuring dog leading the way on our walks. Scent has been healing. It's been transformative. Scent transported her from the lab into a wide-open world of pleasurable experience. But scent can also, still, carry her in an instant back to the lab. A single whiff of a particular odor can open up those affective scars and momentarily undo the years of healing she has undergone. When we visit the vet, she smells the haunting odor of disinfectant long before we get into the building—and what else? The fear of other dogs? The vets themselves? The odor of medications and medical equipment imperceptible to or ignored by my nose? These smells send Saoirse into eruptions of shakes and shivers as we make our way into the vet's office, and she climbs onto my lap as we wait for our appointment. Alexandra Horowitz writes, in her work on dogs and scent, "Smell rubber-bands time for dogs, pulling some of the past and future into the now."[11] Scents, then, can be sites of memory and haunting, bringing the body-mind right back to moments of fear and trauma.

As Saoirse adapted to living with us and with our elder dog, Maizy, she would study Maizy's behavior with a resolute earnestness, learning from her how to be a dog living in the world beyond the lab. She watched how Maizy walked on a leash, rolled on her back in the cool grass, played tug-of-war, ran after tennis balls, played with other dogs, and swam. Saoirse discovered that many of the things Maizy loved were not things she enjoyed. Saoirse watched Maizy leap into Lake Washington with total abandon, paddling around as many Labradors love to do. Wanting to try it out for herself, she stuck one paw in the water, only to dash back up the beach to watch Maizy from a safe distance. She found that she didn't care much for tennis balls or tug-of-war or playing with other dogs. But she discovered without delay that she loved rolling on her back in the grass and racing around the yard, long ears flapping in the wind.

Years after Maizy's death, every time Saoirse rolls in the grass with total abandon, I'm reminded of the joyful presence of Maizy and the way she taught Saoirse how to find delight in herself as a dog and in the world around her. When Saoirse rolls in the grass, I feel Maizy's ghost lingering, her gentle, caring, and loving presence haunting us. Haunting can be characterized as a calling forth of violent histories, trauma, grief, and loss, but it can also be an experience of recalling love, joy, connection, and beauty. Haunting is a relationship or feeling not easily forgotten; it's something that stays with us, coming up again and again, a reminder of the past, a comforting presence when we may need it most.

A year into Saoirse living with us, we got a call from the lab. Over that past year and in the several years that followed, I helped Allison search for adoptive homes for the dogs in the lab. It had adopted out the two hounds Allison had mentioned when I visited the lab for the first time to different friends of mine, and several other beagles had also found homes outside the lab. When Allison called me this time, it was with a slightly different request. One of the beagles who had been adopted out the year before we adopted Saoirse needed a new home. This beagle, Lucy, had been taken in by an older man and woman and had lived with them for a year; then, quite suddenly, the woman died. Her husband was bereft. He was grieving for his wife and for the fact that he knew he couldn't take care of Lucy on his own. He called the lab to ask for help. Allison called me, and we quickly found several different homes interested in taking her, but all of them fell through at the last minute. Finally, without other options, the man and his daughter were going to take Lucy to a shelter in the hopes that she would be adopted that way. I told them we would adopt Lucy instead, and that she would get to live with another beagle from the lab. We arranged to meet in a parking lot at a mall halfway between our homes.

The man hugged me and sobbed as he said goodbye to Lucy. His grief was palpable as he explained through his tears that his wife had loved Lucy deeply and that his heart was broken that he couldn't take care of her on his own. Lucy, too, was grieving, and the two of them together reflected the love and loss they were experiencing.

When we brought Lucy into the house to meet Saoirse and Maizy, Lucy and Saoirse ran to greet each other with tail wags, sniffs, and licks. We had never seen Saoirse act like this with any dog, including Maizy. They were completely at ease with each other right away, immediately settling into a routine of playing together, sleeping together, and moving along shoulder to shoulder on our walks. They always wanted to know the other was close by. As it turns out, Allison told me that they had been housed together in the lab and likely remember each other from that time, reunited in this new world outside the lab.

Although Lucy seemed overjoyed to meet Saoirse and Maizy, she was also clearly depressed when she first arrived in our home—something the vet validated when we took her in for her first exam. After surviving the lab and finding a home, she had lost the person she was most attached to, after which she was taken to another entirely different home with humans she didn't know. The losses she had experienced were significant. Compounding these losses was the fact that Maizy was at the end of her life and died several months after we adopted Lucy. Both beagles were quiet and subdued as we gave Maizy end-of-life palliative care at home and in the days after Maizy died. During the first months of Lucy living with us, and then through Maizy's death, Saoirse comforted Lucy; she would gently lick her head and ears and snuggle against her on the couch. Lucy gradually recovered, growing into a highly social, energetic, and loving dog.

Six months after Lucy had finally settled in, we had a series of beagles pass through our home as an overnight stop on their way from the lab to their longer-term foster or adoptive homes. One of these beagles hit it off with Lucy and Saoirse, piling into the dog pile to sleep, falling into a routine with them and with us within only hours of her arrival in our home. The morning after we picked her up and brought her home, we received word that the foster home that had been lined up for her had fallen through. And so, Amelia joined the pack, making her life with Lucy and Saoirse, and with us.

Where Saoirse was shy, fearful, and easily stressed, Lucy was the opposite—laid-back, outgoing, and playful with those she encountered (dog, human, cat, squirrel, or crow). Things that terrified Saoirse didn't faze Lucy. Where Lucy was easygoing and friendly, Amelia was circumspect and feisty with strangers and the vet, and intensely affectionate with us, seeking comfort and contact from us and from the other beagles almost constantly. Over the years, these qualities have both shifted and deepened. Saoirse has grown more confident, not so easily frightened, and bolder with people she doesn't know. Lucy has remained consistent in her outgoing, sweet, and playful personality, but as she has aged her anxiety about being apart from me has deepened and she follows me around the house, laying her head on my feet at my desk, as she is doing now, as I write this. Amelia's affectionate nature has intensified, her need for one-on-one care and connection a powerful force. When I arrive home, she'll wait while I crouch down and greet the other dogs, and then she'll stand on her hind legs, putting her front paws on my chest, bowing her head for ear scratches and kisses on her forehead. She will stand like this for as long as I have the patience to kneel in front of her.

Saoirse's, Lucy's, and Amelia's life experiences, their responses to them, and their personalities are, of course, their own—they are unique individuals. But they have also been shaped by their breeding. Beagles are the most widely used breed of dog for biomedical research because they are docile and easy to handle, rarely bite, and are small and thus more affordable to cage and feed. Breeding is a powerful technology of control, intervention, and shaping of animals' lives and bodies to enable easier instrumentalization by humans. And it's a form of control that is largely invisible, as normalized ideas about particular species and breeds become taken-for-granted knowledge (e.g., the idea that cows just naturally give copious amounts of milk—a belief that obscures the long history of intensive selective breeding for increasing volumes of milk production for human consumption). All domesticated animals today are the result of many generations of breeding that relies on the oversight and control of their reproductive lives.

Many pure dog breeds are bred to adhere to aesthetic standards determined by breed organizations, even as these desired qualities can have detrimental effects for the dogs themselves (as in the case of the English bulldog, for instance, whose selective breeding for a flatter face and muzzle has caused severe respiratory issues, and whose aesthetically desirable deep skinfolds are prone to infection). The focus in breeding on ideas of purity and strategic crossbreeds (as in the case of Labradoodles, bred to introduce hypoallergenic qualities into the Labrador Retriever), in addition to the very fact that breeding exists as a common practice at all, reflects the prioritization of human interests and expectations about what a particular animal will be—what they'll look like, what their body will be and do, what their personality will be, and what they will be used for.

As a breed, beagles have been primarily maintained throughout history for hunting, an activity they are well-suited for with their powerful sense of smell, their loud howl to alert hunters to the presence of an animal, the white tip on their tail that is easy to spot in tall grass, and their enthusiasm for chasing small animals. Beagles, like other purebred dogs, are also prized for their aesthetic qualities; in fact, beagles have won Best in Show at the prestigious Westminster Kennel Club Dog Show twice (Uno in 2008, and Ms. P in 2015), resulting in cult followings for both of those specific beagles, and for beagles in general. In 2015, in the months following Ms. P's win, people in our neighborhood would go out of their way to come and say hello to the beagles on our walks, rhapsodizing about how they were a prizewinning breed with wonderful personalities and handsome looks, and exclaiming about how lucky we were to have three. These people were shocked when I told them these beagles were used for biomedical research, and I explained that the qualities that make them good "pets" also make them good research subjects. Breeding beagles as pets and for research can be understood on a continuum of how and why we value particular breeds, like beagles, and how and why breeding as a technology of the management of life is sustained in dogs and so many other animals.

In the context of laboratory research, breeding passivity and other desirable qualities for research into the beagle breed is only the first part of crafting the ideal research subject. Also important is the role of care as a mechanism of control, with animal management practices in labs "actively supporting instrumentalization."[12] Writing on the breeding and care of beagles in lab settings, Eva Giraud and Gregory Hollin explain that "care is not always in negotiation with or opposed to instrumental forces. In certain contexts care is precisely what *enables* the instrumentalization of life, in being used to gain knowledge about entities that can be exploited for the purpose of control."[13] In many breeding facilities, socialization with other dogs and with humans is a key practice employed to pacify resistance in the dogs and shape their behaviors so that they more easily comply with their role as research subjects. In the design of the "laboratory beagle," researchers "were learning from [the dogs] in order to *actively manipulate* these [desirable] qualities, moulding the animals into 'experimental dogs': the researcher 'having kindness' greatly improves the qualities of the beagle within a laboratory setting."[14] Welfare considerations in breeding and lab settings, then, are at least in part oriented around ensuring the beagle is sociable, easy to handle, and amenable to being poked and prodded without biting or struggling.

As my interest in the experience of animals in biomedical research grew, I wanted to learn more about institutions that routinely use them for experimentation and testing. This took me to the Institutional Animal Care and Use Committee (IACUC) at the University of Washington. These committees are meant to help track and ensure compliance with welfare standards, and they meet regularly to review research protocols, address welfare issues and violations, and consider other questions that arise in the care and use of animals in research.[15] Because IACUC meetings in many states are required by law to be open to the public, for those interested in understanding how these committees operate, these meetings offer a first-hand glimpse into the ethical deliberation and oversight that shape the lives of animals like Saoirse, Lucy, and Amelia during their time in the lab.

The first time I attended an IACUC meeting, I thought I'd come to the wrong place. I peered through the narrow window in the door to see a tiny classroom. I poked my head in and asked if this was the IACUC, and a man setting up a laptop and projector confirmed that it was. As soon as I was settled, I realized the committee was on video projected onto the screen at the front of the room. The committee members were seated around a horseshoe-shaped table, in an undisclosed location, presumably somewhere on campus. The only other two attendees in the room were individuals I recognized as dedicated members of Seattle's animal advocacy movement.

This separation of the public from the committee causes these meetings to have a disembodied quality. It struck me that my experiences with Lucy, Saoirse, and Amelia have been ones of intensely embodied interactions—the feeling of Saoirse shaking in fear against me, the scratching of their claws on my legs as they jump up to say hello when I come home, the warmth of their bodies and the feeling of their breath on my face as we pile in bed together. These embodied connections and the emotional experiences with which they are connected—the way that the beagles themselves are embodied beings in relationships with each other, with me, and with the world—were, as it turns out, considerations missing entirely from the IACUC's deliberations.

Today, the remote meeting format doesn't seem so odd since the COVID-19 pandemic rendered many meetings remote, but in the years preceding that global pandemic, it was noteworthy that the committee and members of the public were never in the same place. These meetings were not always characterized by this level of separation, however. Following an unrelated arson on campus committed by the Earth Liberation Front in 2001, the University of Washington's animal research sector became increasingly anxious about animal rights activists who were peacefully protesting the use of animals in biomedical research, and they shifted the meetings to video to avoid face-to-face interactions with activists and other members of the public.[16] When COVID-19 drove many activities fully online, the IACUC meetings (including the public's participation) were conducted entirely on Zoom.

The activities of the IACUC, visible in small part in these public meetings, are meant to monitor research activities *internally*. *Externally*, the USDA Animal and Plant Health Inspection Service (APHIS) monitors animal care and use in research settings and conducts routine surprise inspections of the labs to ensure animal welfare standards are being met. I spoke with one member of the animal care staff at the University of Washington who reported to me that these inspections are often not "surprise" at all. They explained that APHIS will conduct inspections of all research facilities in Seattle at roughly the same time, and there is an unspoken agreement across these institutions that they will give a heads-up to other facilities in the region when the inspector visits the first facility. The stakes of adequately reporting and responding to animal ethics violations can be high—these violations can reach the public realm (USDA inspection reports are public record) and may harm the reputation of the institution or even result in legal action against a particular researcher or the institution itself. In cases where issues arise, administrators and the IACUC often try to either minimize the problem or resolve it as quietly as possible.[17]

The IACUC meetings are one place where information on noncompliance—termed *adverse events* (where there is an issue with the care or housing of the animals that rises to the level of violating minimum animal welfare requirements)—comes to light for those who know to look for it there.[18] In University of Washington IACUC meetings, one of the routine announcements is a reporting of the adverse events that occurred that month—for instance, month after month a recurring "adverse event" was the repeated failure of staff members to check that the automatic watering and feeding devices in animals' cages were working correctly, resulting in the regular deaths of rats and mice from dehydration and/or starvation. These deaths are usually met with reprimands by the IACUC in the form of a letter to the lab in question, and/or requirements for retraining staff on animal care. However, they continue to occur, highlighting the inadequacy of even the most basic animal care provision and the total precarity of animals who live and die in labs—who represent "the paradigm example of vulnerability."[19] Captive in cages inside locked rooms inside locked hallways inside highly securitized facilities, animals have no way to satisfy their basic needs. They are wholly dependent on humans for their care and survival, and wholly vulnerable to the ways humans treat and use them.

Based on my experience attending many of the University of Washington's IACUC meetings, I noticed a pattern in the persistent absence of any consideration of the actual experience of animals and the lack of attention to any ethical questions related to animal research.[20] Historically, the IACUC deliberated over proposed protocols in the public meetings so that there was some level of transparency about, one, the kind of research that is being approved, and two, the ways these decisions are discussed and made. Decision-making about whether a protocol would be approved (and it is *extremely* rare for any protocol to be denied) is now done entirely offline, with no transparency for the public about what kinds of research public tax dollars are funding and how animals' interests have been considered in the process. Today, these meetings, which when I began attending them lasted an hour or more, have been reduced to roughly twenty minutes and reflect the perfunctory nature of the committee itself and a general disinterest in addressing animal welfare with a base level of public accountability.

One of the most important questions to ask in reflecting on both the nature of the IACUC and the use of animals in biomedical research more generally is whether welfare of animals in these settings can exist at all. If welfare means *to fare well*, can we consider animals to be faring well when even the most basic forms of autonomy over their lives and relationships are, by the nature of their captivity and use, curtailed entirely? Welfare is also defined in an abstracted way—certain basic qualities of living conditions and social interactions are defined for species, but the nuances and specificity of individual animals and their experiences are missing from these definitions or how welfare practices affect their individual lives.

The lives of animals in labs, like Amelia, Lucy, and Saoirse, are thoroughly shaped by human interests and dictated from birth to death by human decision-making oriented around the animals' usefulness to science. In addition to the pain many animals experience in research protocols, captivity, boredom, isolation from others, perpetual enclosure in indoor spaces, impoverished living conditions, and breeding programs are just a few of the routine qualities that characterize the lives of many animals in lab research. John P. Gluck, a former primate scientist and animal ethicist, recalls his work with animals in labs: "Although I had ensured that my animals were 'cared for' in accordance with existing standards, I created for them an existence in which their attempts to express their basic natures were more or less thwarted at every turn. They led stunted, unfulfilled lives of boredom punctuated by episodes of fear and pain. Or worse."[21] For the same reason that laws protecting vulnerable research subjects (human or animal) are problematic, IACUCs disguise the underlying harm done to those most vulnerable, dividing procedures and practices into those that constitute *acceptable/necessary* and *unacceptable/unnecessary* harm. But these measures rely on a normalized acceptance that a certain level of harm is necessary for the "greater good" (*for humans*) and that certain categories of life are always going to be "sacrificed" for the greater good of those in positions of power and privilege. As Gluck explains:

> Most scientists—even those who purport to have the best interests of animals in mind—still hold fast to the presumption that animal use is justified a priori because it yields considerable benefits for humankind. The harms to animals that flow from this stance, abetted by career ambition and willful blindness to the actual welfare consequences of animal acquisition, captivity, and experimentation, remain considerable. . . . Being an animal researcher is not just about theory construction, scientific methodology, and data collection, but about being the creator, purchaser, and terminator of lives. Lives that have, in the broadest sense, meaning for the creatures who are living them, independent of the goals that we as researchers have for them.[22]

These forms of meaning are eclipsed by their use in research, and the costs of the routine denial of animals' experiences are high. My years living with Saoirse, Lucy, and Amelia have shown these costs in stark detail, thrown into relief even more clearly by witnessing their radical transformation as they discover the things that have meaning to them and as they create lives of flourishing within the bounds of our home.

Animal welfare laws—however lax and ineffective—aim to address what's needed for species like dogs and nonhuman primates to have better lives in the lab. But what about the nearly fourteen million animals used in the United States each year who aren't covered by the Animal Welfare Act?

Nonhuman primates and dogs are the charismatic megafauna of the animal research world. In addition to being among the species protected by the Animal Welfare Act, they are also the species most widely recognized, identified with, and worried over in the public sphere. Lab staff I spoke with explained that they were eager to have dogs and nonhuman primates phased out of research. At first, I thought they were talking about phasing out animals more generally, but it became clear that they were talking only about dogs and monkeys. They clarified that these animals should be replaced with other species—rats, mice, amphibians, or fish—with whom they identified less.

Decisions about which species are used follow a hierarchical logic of species' perceived capacities and lifeworlds, and foundationally which species humans relate to and whose welfare and suffering they care about most. In this formulation, nonhuman primates sit at the top of the hierarchy, and a replacement strategy would involve replacing the monkey (higher order) with a dog (lower order), or a dog (higher order) with a rabbit (lower order), or a rabbit (higher order) with a rat (lower order), or a rat (higher order) with a fish (lower order). As this illustrates, these designations of higher/lower order rest on shifting terrain that hierarchizes life in and beyond the lab.[23]

Rats and mice, as so-called lower-order animals, have long been important species for biomedical research and preclinical trials because they have similar physiology, anatomy, and genes to humans; they are small, easy to reproduce in large numbers, and easy and inexpensive to feed and house; and they have short life cycles. But who are these rats, their lives and relationships obscured from those outside the lab? Who are they to themselves, to strangers, and to their kin? Marc Bekoff, a behavioral ecologist and ethologist, summarizes some of the recent research on the cognitive and emotional capacities of rats:

> Detailed research clearly shows rats cut deals and trade different favors with one another (they follow "tit-for-tat" when trading grooming for food, and food for grooming); know when they've forgotten something; dream of a better future (just as humans do); display empathy for other rats by reading pain in their faces; save drowning rats rather than eat chocolate; play, laugh, and like to be tickled; tell you they're happy by relaxing their ears; regret what they didn't do and recognize what might-have-been; and free familiar trapped rats from being restrained.[24]

The qualities of who they are as individuals and collectives are not guiding factors in using rats in research—in fact, these qualities are routinely denied in the ease with which they are instrumentalized for research.

Like rats and mice, zebra fish are one of the top species used in laboratory research worldwide. Zebra fish reproduce prolifically (producing two to three hundred fertilized eggs per week), and their eggs are transparent early in their development, which offers researchers the chance to study genetic development.[25] They are used in research on conditions that humans experience, such as cancer, tuberculosis, schizophrenia, and muscular dystrophy—diseases researchers study by injecting genes into the embryo that cause these conditions to manifest in the fish.[26] Zebra fish have rich emotional inner worlds. Because zebra fish are useful in understanding human brain disorders, researchers have studied in depth both cognition and emotion in the species. Zebra fish experience fear and anxiety: "Like humans and other mammals, their future behavior can be rendered maladaptive by extreme stress . . . reporting impaired learning following acute predator exposure."[27] The recognition of emotion in other species *can* be a window into empathizing with them, and animal behavioral studies can offer us information about how animals experience emotion; however, the costs of studying animal emotion are often high. For instance, the study just mentioned that states that extreme stress can impact their future behavior involved early exposure to an electric shock that the fish could not escape. Realizing that the shock was inevitable, the fish experienced a "learned helplessness [which] is a validated model for depression."[28]

Writing on the depression animals experience in research, often as a result of the conditions humans deliberately create for them, the philosopher Chloë Taylor explains that "scientists invoke depression in lab animals in order to learn about the causes and possible treatments of depression in humans, and thus know that animals can suffer from despair in ways similar to humans."[29]

Zebra fish, rats, mice, and other species used for human biomedical research occupy a curious conceptual place within the medical and scientific communities wherein researchers accept and deny the similarities between humans and other species and determine in what context these similarities and differences matter. When it counts, in terms of justifying the use of animals in research, a wide range of other species are framed as being extremely close to humans in their physiological, psychological, and social systems. However, when confronted with the ethical inconsistencies of using animals for invasive and harmful studies (studies that, today, would likely be prohibited as unquestionably unethical if the research subject were human), animals are framed as being dissimilar enough from humans to warrant different, less rigorous forms of ethical consideration.[30]

I bring these animals (designated as *lower order*) into this story to make them present, to honor their presence even in their absence. But what I can offer here is an at-best-impoverished glimpse of these species, and especially the lives of any of the individuals of these species. I've never lived intimately with, gotten to know, or spent time with any individual ferrets, mice, zebra fish, or many of the other species used for research. Instead of the richer understanding of the impacts of biomedical research on the beagles with whom I share a home, these other species are like ghostly traces haunting the pages of this book, whispers of lives lived in cages and consumed, never materializing in this text in corporeal form.

In reckoning with the kinds of lives animals live in laboratories, and the ethical ambiguities of using animals for research at all, we come up against calculations that weigh human life and well-being against the lives and well-being of animal life in biomedical research. Although there are many instances where hierarchies of human supremacy and exceptionalism sustain routine and normalized harm, there are few places where the calculation of human life versus animal life is so clearly articulated. In 2011, the Foundation for Biomedical Research launched a billboard advertising campaign in select US cities that set up this dichotomy (human life versus animal life). On the left side of the billboard is a white rat of a breed commonly used in biomedical research; on the right side is a young, white, blonde girl with a big smile, cradling her chin in her hands. The text joining the two images reads, "Who would you RAT/HER see live?" (the "rat" or "her").[31] Frankie Tull, the president of the Foundation for Biomedical Research, explains: "Our new billboards ask people to consider an important ethical dilemma we face as a society: Would you rather do away with animal research or have the new medical cures, treatments and therapies for which so many people desperately wait?"[32]

There's a lot going on in this statement. The very existence of this billboard acknowledges that there's an active public debate about whether animal research is justified, and that the shift in this public debate has been significant enough to warrant the research community spending money on advertising to reinforce the narrative that animal research is necessary. The underlying presumption is that animal research *does* result in cures or treatments to human disease—the accuracy of this claim is not brought into question; rather, the ad campaign reinforces the taken-for-granted idea that animals must be sacrificed to develop these important human treatments.

Who is represented in this billboard is also telling—the foundation has chosen a little girl who is white and blonde and has a decidedly innocent look, her chin resting in her hands, smiling sweetly. It is not a coincidence that the child represented is blonde and blue-eyed—whiteness, and particularly an Aryan type of whiteness, represents the pinnacle of humanity in Western Enlightenment formulations of the human. It is also not a coincidence that she is a little *girl*, her femininity adding to the onlooker's gendered understanding of her vulnerability. The animal they have chosen is a white mouse with red eyes—another poignant choice. It's not a little girl pitted against a cute beagle puppy or infant primate. Although the answer to the question of who you would rather see live would likely be the same for any animal species juxtaposed with the child. The obvious answer for most people would be to choose the human child without a second thought, which in and of itself reflects a sense of human supremacy. But a dog or a primate, who might evoke more empathy and doubt than a rat, might also have the unintended effect of alerting the public to the widespread and perhaps less palatable use of both nonhuman primates and dogs in medical research.

I've spent a lot of time thinking about the medications and medical interventions that have allowed me to survive and to live a relatively healthy life. I've been a lifelong sufferer from asthma. As a toddler, I was hospitalized for an extended period when my asthma was exacerbated by pneumonia and became life-threatening. Later, periodic visits to the ER for emergency breathing treatments just felt like a learned reality of being an asthmatic. It's terrifying to not be able to breathe. The instinct in these moments is to take great, gasping, gulps of air to try to fill your lungs with oxygen. It's completely counterintuitive to have to learn as a young child to ignore that instinct to gulp air, and instead take tiny, shallow sips of air during a bad asthma attack until you can access medication that will make it possible to breathe again.

The nature of privilege is that it seems to operate invisibly for those who benefit from it. I didn't think of the medical treatment of my asthma in terms of the inequitable access to medical care along lines of race and class, and I didn't think much about the medication I needed unless it wasn't available at the moment I needed it. I didn't think about this medicine as a commodity made more affordable by the health insurance my family possessed, or as the product of a long history of biomedical research.

And I certainly didn't connect it to the countless animal lives used in this research so that I and so many other humans could have access to necessary medications. Ventolin HFA, Advair Diskus, Symbicort, and Flovent HFA are common prescription medications used to treat and/or prevent asthma attacks and bronchospasm—all of which I have used at one point or another to control my asthma. These drugs represent four of the top twenty-five most utilized prescription drugs on the market developed through animal testing.[33] The Foundation for Biomedical Research reports that Ventolin HFA, as number 4 on this ranked list, was developed using rats, beagles, and New Zealand white rabbits; the development of Advair Diskus (number 5) relied on rats, dogs, mice, rabbits, guinea pigs, monkeys, and hamsters; and Symbicort (number 22) and Flovent HFA (number 24) used mice, rats, and dogs.[34]

As I studied the connection between the use of beagles and other animals for research and my asthma medication, I was sitting on the couch, Saoirse curled up against me, snoring loudly; Amelia slept on the other end of the couch on her back, her legs sticking up in the air; Lucy snoozed in her bed across the room, waking occasionally to make sure I was still there. My asthma had been particularly bad in the preceding months because of the thick smoke that had settled in Seattle from the wildfires in California, Oregon, Washington, and British Columbia. I was relying heavily on a combination of inhalers and oral medications to get through each day, and still, I was barely controlling my asthma. I depend on and am grateful for my asthma medication and for the other medications that I and others take to both sustain and enhance the quality of our lives. However, I can't help but wonder, *at what cost?* I thought even about the phrase *taking medication* as a provocation, an implicit acknowledgment that something is being taken in its procurement—*from whom* is left unsaid. Faced with the lives of these three sleeping beagles, it was impossible to deny those from whom these medications were taken. These were a lucky few of the small number of animals who ever make it out of the lab; the overwhelming majority of animals used for research die as a part of the studies for which they are used, are euthanized after the studies are completed, or live out longer lives in the confinement of the lab setting.

Several years after I started working with Allison to find adoptive homes, she quit her job. I had been meeting semiregularly in the preceding couple of years with Dave Anderson, the director of animal research at the University of Washington, to try to arrange the relocation of pigs from the labs to Pigs Peace Sanctuary in Stanwood, Washington, and primates to a sanctuary in the southwestern United States that had agreed to take them. Although the university had gone to the trouble of visiting these sanctuaries and approving them as locations for the "retirement" of pigs and macaques, it never identified a single animal who was what it considered an adequate candidate for removal from the lab.

When I learned that Allison was quitting, I immediately scheduled another meeting with Dave to discuss how the adoptions of dogs and other animals would continue. He said that there was no one to take over handling the adoptions—that Allison had done this, uncompensated, on top of her other duties; that there was no one else willing to take on this level of unpaid labor in the labs; and that there was no funding to compensate someone for managing the adoptions. I asked if this could be taken on by a trusted volunteer and offered my own labor for the role. He said that would not be possible. I asked what would happen to all the dogs and other animals for whom attempts would previously have been made to find adoptive homes. He said they would all be euthanized unless a researcher or veterinary staff member chose to adopt an individual animal.

Now, when I look at Saoirse and Lucy and Amelia, I see not only their own histories in the lab and the countless others who are used and used up for research but also the ones who might have gotten out. The fact that some dogs and members of other species had been adopted out at all does not justify the continued use of animals in research, although from researchers and veterinary technicians I talked to, I heard that it makes many in the research and animal care community feel better about the costs of animal research and accept the harm they witness. I wondered if adoptions were adding to a culture that continued to normalize the violence of experimenting on animals in a way similar to notions of welfare in the lab do. However, it's hard to look at these three beagles whose lives are so entangled now with each other's and with my own and imagine them not having the opportunity to leave the lab, to be haunted by their lasting trauma, to heal, and to live.

In reading and thinking about animal research, I've wondered what motivates calculations of certain lives weighed against others, like humans against other animals. And I wonder to what extent animal research actually results in solutions to humans in need of medical interventions. What is it that sustains these cultures of harm and the disavowal of the at-once disturbing and mundane nature of this harm? The use of animals in biomedical research is overwhelmingly justified by the potential benefit it has for humans—promises of lifesaving new drugs, procedures, and technologies that will extend people's lifespans and improve their quality of life. But a question fundamental to this belief is whether these promises are fulfilled *by the actual use of animals*. This question is obscured and overlooked for several reasons, not the least of which is that the animal research industry is big business. The political economy of animal research involves a complex architecture composed of researchers and the institutions that employ them; both public and private funding agencies; government agencies that regulate animal research and require animal testing to be used in advance of the approval of drugs coming to market; animal breeders and producers of caging, food, and other animal care infrastructure; and corporations (e.g., pharmaceutical companies) that ultimately develop, market, and sell the products developed through animal experimentation.

The significant allocation of research funding earmarked for animal research is a major factor that propels the ongoing use of animals in research protocols. The National Institutes of Health (NIH), one of the major funders of medical and scientific research in the United States, distributes $5.5 billion annually for animal research.[35] I spoke to one scientist who told me anecdotally that she had met some researchers who will actually write animals into their research protocols when they don't need to be there simply because funding for research projects involving animals is so abundant and the inclusion of animals could help them get their project funded. But the availability of research funds and the normalization of animal research through these funding streams can obscure the harm caused to the animals themselves.

Andrew Knight, an animal ethicist and veterinary specialist, argues that the costs to animals of biomedical research do not justify the benefits and that public declarations by the scientific community about the value of animal research (like the Foundation for Biomedical Research billboard) grossly overstate the benefits of animal research for humans.[36] In many cases, these claims of the success of studies involving animal research cannot be substantiated, or they are validated with false or skewed data.[37] Studies in which animals were used and which were not successful in achieving their goals are typically not published, and so there is a disproportionality in reporting findings wherein *negative findings are not counted*.[38] Scientists may also publish only partial findings from a study—the parts of the study where their goals were achieved and not the instances where they had negative findings.[39]

The low rates of efficacy and high rates of failure in developing drugs and treatments that actually work in humans after success in animals are increasingly acknowledged by those in the medical research community. The Government Accountability Office notes that "in recent years, researchers have reported low success rates in reproducing and translating the results of animal experiments. Reproducibility of a study can reflect how reliable its results are, and translatability enables animal research to benefit human health."[40] Joseph Garner, an associate professor in comparative medicine at Stanford University, writes, "Every drug entering human trials, by definition, 'worked' in an animal model in terms of both safety and efficacy, and efficacy is the primary reason why drugs fail in human trials. Thus, the primary reason for these failures can be traced back directly to false positives in animal models committing the pipeline to develop a drug that will ultimately fail."[41] And he continues: "As researchers we assume that our work is meaningful, but the simple fact that roughly 90% of compounds entering human trials will fail should be a harsh reminder that, as [Stanford professor of medicine John P. A.] Ioannidis bluntly pointed out, *'most published research findings are false.'*"[42]

Even Elias Zerhouni, a former director of the NIH, acknowledges the inefficacy of animal research. An article by Rich McManus relates comments made by Zerhouni during a scientific review management board meeting in 2013: "'We have moved away from studying human disease in humans,' he [Zerhouni] lamented. 'We all drank the Kool-Aid on that one, me included.' With the ability to knock in or knock out any gene in a mouse—which 'can't sue us,' Zerhouni quipped—researchers have over-relied on animal data. 'The problem is that it [animal research] hasn't worked, and it's time we stopped dancing around the problem.... We need to refocus and adapt new methodologies for use in humans to understand disease biology in humans.'"[43]

There is much to unpack in Zerhouni's statement here. Aside from the obvious—that he is asserting the inefficacy of animal research—he also reveals quite a lot in his statement that mice "can't sue us," wherein he implicitly acknowledges that animal research involves a level of such significant harm that, if delivered to human participants in research studies today in the way it is delivered to animals, would warrant litigation. He also brings up a perennial and serious ethical quandary in lamenting the fact that researchers have "moved away from studying human disease in humans." With the historical and contemporary exploitation of humans in vulnerable positions for biomedical developments, returning to a more wholesale use of humans in early medical trials is highly fraught and raises profoundly troubling ethical questions. But perhaps the answer is not to sub in the extremely vulnerable cohort of nonhuman animals to avoid reproducing the great harm posed to human research subjects. Perhaps the questions we should be asking are different. *Is the infinite pursuit of new biomedical innovations justified if it is premised on profound and irreconcilable harm to those who are most vulnerable among us?* What are the underlying structures of violence that make it acceptable to steal millions of nonhuman lives from them each year in the hopes that humans might benefit? What is it that makes animals' lives so valuable in research settings and so unvalued as subjects deserving of autonomy over their own bodies and lives?

As I researched questions of efficacy in animal research, I discovered that, although animals were used in these studies, even my asthma medications were not successfully developed because of animal testing. In fact:

> Animals are poor candidates for studying asthma because the anatomy, immune system, and inflammatory responses exhibited by animal lungs differ greatly from those in humans. Animals used to study this condition do not exhibit symptoms similar to human asthma—asthma is a human disease—, so typically the disease has to be artificially introduced in the airways. Animals also have different airway architecture and different breathing behaviors that affect where the inhaled irritant "lands" in the lung and therefore how the organism reacts. Furthermore, the distribution of lung inflammation is different, and many animals become tolerant after repeated allergen exposure. Therefore, key features of human asthma cannot be recreated in animal models.[44]

Does this mean that those countless animals (the mice, rats, guinea pigs, hamsters, monkeys, rabbits, and dogs) who were bred, used, and killed for asthma research—who were held up as success stories and as justification for the continued use of animals in research—experienced impoverished lives of suffering, harm, and isolation for nothing other than an abstract perceived benefit to sufferers of asthma and the funding and accolades bestowed on the researchers who used them?

If animals are so readily available, and animal research so taken for granted and entrenched in funding streams and academic professionalization and recognition, the motivation for developing alternatives is very low. Although nonanimal models for studying human disease are slow to develop, there has been promising progress made in technological innovations that more accurately replicate human physiology and disease. In the context of asthma and lung disease research, for instance, using their organ-on-a-chip technology as a model, Harvard researchers have developed a small airway on a chip to re-create a tiny model of the living human lung that can replicate common lung diseases like asthma and chronic obstructive pulmonary disease (COPD); they then can test the efficacy of different medications and treatments on those diseased lung models.[45] There have also been significant advances in AI and machine learning that show promising results for studying the effects of therapeutics on human organs and biological systems, and the rapid evolution of these technologies could be used to conduct more effective treatments for human disease.[46]

Although there are advances like these and others, animals are still widely used in developing and testing new vaccines and drugs, and technologies that provide alternatives to animal research have not been as widely developed or adopted as they could be. This innovation has been inhibited significantly by the lack of funding available for exploring and implementing these cutting-edge technologies and by entrenched cultures of normalized animal use. Were the funding, time, energy, and brilliant minds dedicated to research involving animals to be focused instead on developing more accurate and effective models of research designed to replicate human biology, perhaps we would have a much larger and diverse tool kit of alternatives to animal research that could provide more advanced treatments and techniques for treating the diseases that plague humans.

This focus might be possible if at root there was a wider acknowledgment of the serious limits of animal research in terms of delivering actual outcomes. And it also might be possible if we were to engage in a deeper interrogation of our (humans') sense of entitlement to the bodies and lives of others. Attentiveness not only to the embodied forms of harm animals experience as a result of this entitlement but also to the emotional and psychological hauntings of these experiences prompts important questions about how we might revoke this sense of entitlement, this easy eclipsing of the abstract idea of human well-being (not even supported by scientific outcomes) over the very lives and existence of other species. To be faced with these individuals, as I have been with Lucy, Amelia, and Saoirse, necessitates this kind of attentiveness. And hopefully, I think, it opens doors to consider those—even those so routinely dismissed as lesser forms of life, like rats and zebra fish—who are brought into being, live, and are killed in labs, individuals who will never see the sky, or feel the damp earth beneath their paws, or catch the scent carried by a breeze to their nose.

How might we undo this entitlement and respond to the irreconcilable harm at the heart of this calculus of human versus animal life? What would it look like to stop appropriating animals' lives and bodies for human projects of progress and innovation and focus instead on the ways we can support each other's flourishing as we follow our own paths and journeys to well-being and care?

It was late afternoon in the middle of winter in the Connecticut woods. The cold day was warmed slightly by the sun, and I inhaled the rich earthy aroma of fallen leaves decomposing on the ground. As usual, the beagles' noses were to the ground, their tails up, following urgently what were probably a thousand different scents down the forest path. I heard a rustle in the leaves and looked ahead to see a brilliant orange fox streak across the path in front of us. "A fox!" I thought, transfixed by the sight. The beagles all at once caught the fox's scent, and their noses came alive even more than before, their deep inhalations audible to me, walking behind them. They pulled in that direction, never looking up to see the fox run across the path. This activation of their sense of smell read to me as though they too were exclaiming silently to themselves, "A fox!"—transfixed by the scent.

Our experience of encountering the fox, shared intimately through such different sensory mediums, jarred me in that moment, as I thought about the radical alterity of our senses—maybe not so much the difference in the senses themselves but in how they dominated and shaped our worldviews and our everyday encounters with the world around us. I thought about how different we were as species and as individuals—not in a way that necessitated a hierarchical ordering of me (as human) over them (as dogs) but instead in a way that honored their animality—their canine-ness, their beagle-ness—and my own animality, my humanness—as together we found meaning in that moment.

I was transported for a moment back to the lab and back to thinking about the hierarchies that are sustained and cause harm. And I imagined how these hierarchies might be dismantled through attending to sensory, embodied experiences that anchor us to the present world around us and to each other in complex entanglements of care. Here we were together on shared and separate journeys that could unfold simultaneously and without causing harm to the other. What different forms of knowing, witnessing, and transformation, I wonder, are possible through attending to our senses in these moments of encounter—in the lab, in the forest, or in our home?

In that Connecticut forest I thought about what it meant to witness these beagles' scent-driven lives as their noses skimmed along the ground in search of . . . a fox! I thought about their breeding as fox-hunting dogs—was there something deep in their genes that recognized that scent, and was this a haunting—their genetic makeup calling up long-dead ancestors? I thought about their lives in the lab, the range of scents they would have experienced and the feelings of fear or a sense of care these scents might have evoked. I thought about those haunting odors of other animals' lives unfolding in the lab that are still present in those basement corridors and those howls still echoing down sterile hallways. Sensory experience powerfully renders these hauntings real and immediate. It can transport us all at once back to past traumas, triggering memories and feelings that we might rather forget—that indeed we might work desperately to forget, to confine to the past. But the sensory can also pull us out of present struggles and relationships of harm, gently tugging like the first few curious sniffs of a beagle's nose at an instinct for curiosity, a desire for beauty and pleasure, to transform painful moments into something that can be tolerated, relished, adored.

4

Consumed by Desire

Extracting is taking. Actually, extracting is stealing—it
is taking without consent, without thought, care, or even
knowledge of the impacts that extraction has on the other
living things in that environment.

—Leanne Betasamosake Simpson, *As We Have Always Done*

I peered down into the galvanized steel stock tubs at the sea of fuzzy
chicks—a moving mass of pale yellow, black, and orange bodies. They
peeped loudly as they scrambled over each other, piling themselves in the
warmest part of the tubs under the heat lamps. It was February and I was
chilled, even in my winter coat, standing there in the drafty country feedstore. On the cold outer edges of each tub there were a few dead chicks
and others who hunkered down, rocking, with their eyes closed.

"Are these ones OK?" I asked the store owner, a middle-aged woman
whose name was Marcie, as I pointed at the dying chicks.

Marcie walked over, glanced down into the tub, grabbed a nearby
bucket, and deftly plucked the dead chicks from the tub, tossing them in
the bucket. She paused with one of the seemingly dead chicks, noticing
that this one wasn't yet dead, and placed her back into the tub where she
had been laying. The other chicks on the outside edges of the tub—the

ones who were swaying and dozing—barely moved when Marcie grabbed them and tossed them gently to the center of the tub, closer to the warmth. The livelier chicks immediately climbed on top of them, pushing them away from the warm circle of light and heat. They would likely die soon.

As Marcie shuffled the chicks into *dead*, *dying*, and *soon-to-be dying*, I thought about their disposability. Chicks are inexpensive—at this feedstore, they were retailing for $1.75, their wholesale price being considerably lower. I thought about the way that even in life—in the delightful liveliness of these newborn chicks—death and disposability loomed. It was as if an aura of disposability haunted the outer edges of the tub, a deathworld, a liminal space between commodification (of those chicks in the middle of the tub who would be purchased) and disposal (of those who ended up in the bucket).

I felt a twinge of nausea looking at the dead chicks at the bottom of the bucket and at the dying chicks being stepped on and jostled by the others. I reminded myself that each of these chicks was someone. I wondered who each of the dead chicks had been in their short lives and who they would have become had they lived. But it was difficult to focus too long on any one chick—both the ones removed from the tub and the ones who clamored in a fuzzy multicolored mass under the heat lamp. Did it even matter which ones we chose? I felt a flicker of the futility of trying to individuate the chicks in their abstracting mass.

Echoing my thoughts, Eric asked, "How do we choose?"—clearly stressed by what suddenly felt to both of us like a weighty and fraught decision.

We were deciding the fate not just of the chicks we were planning to take home; we were also making a decision that affected the others. Who would die in that tub because we didn't choose *them*? Who would be purchased by someone who would slaughter them after a couple of years of egg production? What if the ones we chose didn't survive? What if we made a catastrophic mistake in caring for them and they died?

Marcie's voice cut through these thoughts with her pragmatic advice: "It doesn't matter. Just grab however many you want of each breed and then grab a couple extra."

"Why extra?" I asked.

"You're likely to lose a couple. Some of them won't survive."

Eric and I glanced at each other, eyebrows raised. Suddenly, what we had intended to be a fun and novel outing to the country to purchase some cute chicks to raise for eggs in our backyard felt like a high-stakes life-and-death matter. It dawned on me with clarity in that moment that the country feedstore was selling someone's life and we were buying that life, and that commodifying a life is no small matter.

But the casual nature with which Marcie spoke about the reality of a certain level of expected mortality made it seem less dire. Somehow the pricing of the chicks, too, worked to downplay the gravity of what—or, rather, *who*—we were buying. I thought about what else I could buy for $1.75, my mind immediately going to the coffee kiosk we had stopped at on our way to the feedstore: a croissant, a blueberry muffin, a sixteen-ounce drip coffee. I thought back to when I used to work as a barista, and how we would throw away the stale pastries that hadn't sold at the end of the day. The café owner just chalked them up as acceptable losses—the cost of doing business in goods with a limited shelf life. The dead chicks at the bottom of the bucket would have been the same—a financially acceptable loss and the cost of doing business in goods with a precarious life. But these weren't just "goods"; these were individual living and dying beings, and I couldn't shake the uneasiness in witnessing their disposability. I wasn't thinking about it in these terms then, but now I understand that what we were witnessing was the logic of commodifying a living being and the disposability of a low-value life, like a chick, that is an important dimension of commodification.

"And then there's always the possibility you end up with a rooster," Marcie continued.

"A rooster?" Eric asked with concern. Roosters were illegal to keep within city limits, and so posed the problem of what to do with a member of the flock who turned out to be a rooster. We knew a couple of people in our neighborhood who, discovering they had a rooster, slaughtered the birds for meat. We were most definitely not going to do that, and we knew from our research that it was challenging to find a rescue group or sanctuary that would take roosters, given the overwhelming number who turned up.

"Sexing isn't a perfect science," the owner explained. "The hatcheries do their best, but it's just not 100 percent."

Chicks at commercial hatcheries (where feedstores source chicks) are typically sexed by professionals trained in the process. A technique called *vent sexing* is common in hatcheries, where workers turn the chicks upside down and squeeze them to expose the vent area to reveal whether male reproductive organs are present.[1] This is a risky process. Done incorrectly, the chick can easily be disemboweled. Whatever the method, though, sexing chicks is difficult, and it's rare to have 100 percent accuracy.[2] Although most chicks who make it to feedstores are accurately sexed as female, the occasional male is overlooked, and consumers might end up with a rooster—a reality they encounter much later, when the chick has grown up and one day begins to crow.

"God, I had no idea this whole thing would be *so stressful*!" I exclaimed.

Marcie laughed. "Oh, it's not, hon. Relax, you're going to be fine," she said warmly.

A sign taped on the wall behind each stock tub identified which breeds were available. In one, there were Barred Plymouth Rock chicks, who were mostly black with some pale yellow and who would later grow striking black-and-white barred feathers, and then there were Buff Orpington chicks, who were pale yellow and who would become a creamy golden color as adults. In the other tub were the Rhode Island Reds, who were orange and who would deepen into a dark reddish brown, and Australorps, who were black and yellow and whose adult feathers would be jet-black with a blue-green-purple iridescence.

"Let's get one of each," I said to Eric. I leaned down and tried to scoop up a tiny yellow chick. They scattered, scrambling to get away from me. Finally, I was able to capture one in my cupped palms, and Marcie was there, waiting with a small cardboard box to contain them. We caught three other chicks, trying to be as gentle as possible as they peeped loudly and tried to wriggle out of our grasp. We added a heat lamp, wood shavings, starter feed, feeder, and a waterer to our order, paid, and turned to leave.

"Just make sure to keep them warm!" Marcie called after us.

I hugged the box of chicks close to me and hurried out to the car. We blasted the heat in the car on the hour-long drive home, and I resisted the urge to open the top of the box on my lap until we were home with the heat lamp set up and their enclosure nice and toasty. In the car, we named them for my favorite nineteenth-century novelists: Emily and Charlotte (Brontë), Jane (Austen), and George (Eliot).

The chicks didn't die. We set an alarm to wake ourselves up every few hours through the night in those first few days to check on them and make sure the heat lamp hadn't gone out or set the chicks ablaze and to reassure ourselves that everyone was safe and well. They lived and grew in an enclosure in our home office for about ten weeks, until they were old enough and the weather warm enough for them to move outside. We spent hours each day with them, holding them, stroking their feathers, and talking to them.

They were social and inquisitive, wandering around the room when we let them out of the enclosure to explore, and pecking at every foreign object to assess what it was—a dust bunny, a cat toy, a sock, a stack of books. They explored their new world through pecking. They bonded with us and with each other, sleeping snuggled together, and would half-*cheep*/half-*cluck* excitedly when we came in the room, hopping up on the sticks we had fashioned as roosts so they could practice roosting before they moved to the coop we had built outside.

I remember when they were in their awkward gangly teenager phase, not long before we would move them outside, and I invited a friend and her new boyfriend over for dinner. I had planned a meal of a whole roasted chicken, salad, and garlic bread. The incongruity of buying and roasting a chicken while we were in the midst of doting on and caring attentively for four young chickens in our home hadn't occurred to me. At least, it didn't occur to me until I got home from the grocery store and was unwrapping the plastic that contained the whole dead body of an anonymous-to-me chicken.

Holding this chicken's remains in my hands, I felt their legs—body parts that I'd previously thought of as "drumsticks" and "thighs." I felt the bird's chest and chest bone—body parts I had previously thought of as "breasts" and, somewhere beneath, the "wishbone." I recalled that when I was a child my parents would save this "wishbone" for my sister and me to break, and we would eagerly hope for it to crack with the larger piece in our favor, which meant that our wish would come true. This bird's body felt just like the bodies of Charlotte, Jane, George, and Emily; the only difference was that this bird was dead and cold, missing their head, and their body plucked of every feather.

I could hear the chickens practicing their still-awkward clucks from inside the room off the kitchen. I felt a sickening sense of dread, and my eyes welled up with tears. Our friends were going to be here in a couple of hours. I tried to compose myself and pulled out a pan, seasoned the bird's remains, and roasted them in the oven. I could barely eat the chicken that evening as we sat around the table.

To make matters worse, my friend's boyfriend stated as the food was being passed around the table that he was a vegetarian and would just stick with the salad and bread. My friend was such an avid meat eater it hadn't occurred to me that her new love interest would be a vegetarian. She hadn't mentioned it, and I hadn't thought to ask. I was mortified. I had prepared a main course that one of our guests couldn't eat. And, much worse, I had taken this whole chicken's life for this one forgettable meal.

It turns out, though, that this meal was far from forgettable. Fifteen years later, my face still flushes thinking about it. There are certain moments that create a rupture where the life leading up to that moment and the life following that moment feel like two separate realities. This was one of those moments. After this experience it was like the world around me shifted. Suddenly animal remains were everywhere—the meat and dairy sections at the grocery stores, the menus and plates of other diners at restaurants, the catered food served at university social functions, even the half-eaten piece of fried chicken discarded on the sidewalk. I began to see them for what they were: remains, bodies, corpses, and not "meat." Remains of someone who had been subjected to a process of breeding, raising, slaughtering, and butchering. Remains that were then consumed, and the absolutely mundane everydayness of that consumption worked to neutralize the fact of what or *who* they were and what had had to occur for their remains to be seasoned, cooked, cut, savored, chewed up, swallowed, digested, and shit out.

There are so many ways to think about eating animals and the dynamics of human exceptionalism that shape everyday taken-for-granted relationships of harm and a sense of entitlement to the enjoyment and sustenance eating animal-derived foods might provide, as well as the entitlement to their lives that supplied me with that enjoyment. It's taken years for me to understand the complex histories and practices that underwrote that moment and the web of multispecies violence in which I and others were and are implicated through our consumption. However, in this moment and in the moments that followed, holding that dead bird who would become dinner, all I had was a gut reaction without knowing anything more than the sense of dread and regret I felt holding that bird in my hands.

Consumption can be understood as an act of eating or buying—these are perhaps the most common understandings of the term. However, *consumption* also means "to do away with completely: destroy," "to spend wastefully: squander," "to use up," and "to waste or burn away."[3] When we consume an animal, we destroy and use them up. Everyday acts of consumption are just those: forms of consumption that happen routinely on an everyday or regular basis. The effects of this *everydayness* are not innocuous. To do something every day or with a high level of regularity normalizes it; it makes it routine, ordinary, and mundane. It becomes second nature, and it's hard to think critically in a consistent way about things that are mundane and ordinary—these everyday acts of consumption—unless there is a rupture of some kind in how these practices are conceptualized and felt. These ruptures might be thought of as haunting—a crack through which our own and someone else's pasts, presents, and futures are made visible in a different light and demand to be witnessed.

What are we haunted by in our consumption of farmed animals? It's possible to glimpse the nature of these hauntings if we dig a bit deeper to find what's beneath the surface in consuming farmed animals. Not all consumption of animals relies on commodification, but in the case of farmed animals, the overwhelming majority of animals are commodified, and this commodity status helps to solidify how and why they are consumable. Exploring this status reveals how reproduction and reproductive capital form the foundation of this commodification. These forms of reproductive control and the capital they generate occur in each present moment in which an animal is bred and are made possible by their ownership and more fundamentally by the property status of animals.

Nearly fifteen years after buying those four chicks at the feedstore, in the midst of writing this chapter, I attended a "poultry" auction in Enumclaw, Washington, to better understand the particularities of the commodification of birds for eggs and meat. I had attended a number of cow auctions a decade before for my research on the dairy industry, as well as a couple of multispecies auctions that sold calves, cows, pigs, goats, and sheep. But I had never been to an auction for farmed birds.

As I did with past auctions I attended, I arrived early to leave time to look at the animals ahead of the sale. I followed the sound of a crowing rooster over to an area labeled "Poultry Check-In" and watched as a crowd of people holding cardboard boxes in their arms or stacked on wheeled carts waited in front of the check-in table to register the animals they were going to sell. Most of the boxes had jagged holes or slices taken out of the cardboard for air holes. Some of the boxes were silent and still; others shook with the struggle of the bird or birds inside, accompanied by loud bawking, crowing, honking, or quacking. The back hallways leading into the auction ring were lined with tall shelving units on wheels stacked with boxes of birds. I walked into the auction ring, where there were more shelving units filled with boxes. *How are you supposed to see who you're buying?* I wondered, as I tried to peer through the tiny openings in the boxes. In the other farmed animal auctions I had attended, you could walk into the pavilion behind the auction yard to look at the animals for sale before they were led through the auction ring and sold. Here, there appeared to be no viewing of the birds beforehand.

I found a good seat in the bleachers and waited while the auctioneer (an older man wearing a cowboy hat) got settled alongside a young woman whose job it was to record sales on the computer. There was a table in the middle of the auction ring with two women setting up a line of boxes across the table. Boxes would flow in a line on and off the table as the sale unfolded. As each animal sold, a "runner" would carry the sold box out of the ring to the back hallway and would wheel in a new shelving unit when the ones in the ring were getting low on boxes.

I noticed while I was waiting for the auction to start that about half of the boxes were Amazon boxes, an irony that crystallized the underlying logic of mass consumption that dominates US American culture. Although Amazon certainly did not catalyze mass consumer culture (this was underway many decades before Amazon was even germinating as the seed of an idea), the company has dramatically changed the face of US and global consumption and consumerism. That something can be ordered, shipped, and received in one or two days, or even on the same day, not only has accelerated the speed and volume with which people consume but also has intensified the nature of our expectations about the instantaneous fulfillment of our desires. Amazon is built around "fulfillment" centers—a name that signals this connection to desire, and the sense of fulfillment (no matter how brief or superficial) that material consumption promises. The speed at which you can open an app, click a few buttons, and have a product on its way to you helps to eliminate time to ruminate on whether or not this purchase is necessary or even whether it's something you truly want. A fleeting desire becomes an anticipated reality in moments. It's possible to order something before you've even registered what you're doing.

In a different way, the auction manufactures desire and the speed at which consumption can occur. Sales unfold often in under a minute, and there's a rapid-fire back-and-forth between bidders as an employee in the ring shouts and points, driving the enthusiasm for the animal on sale. This frenzied bidding is sometimes so heated that the animal being purchased becomes incidental to the exchange, surpassed by a much more powerful motivator: the act of consumption (and competition) itself.

I was reflecting on this, remembering my earlier research on auctions, when the auctioneer called for the sale to begin. Immediately, I was surprised to learn that the birds were removed from the boxes in real time. The women grabbed each bird in the boxes by the base of their wings and then, if there was more than one bird in a box (as most boxes contained multiple birds), would grab the other birds, loading up their hands with as many as they could hold, crushed together, their wings held back and their bodies often struggling until they realized they couldn't get away and gave up. The handling of the birds was surprisingly rough, but this wasn't unusual or unacceptable within animal agriculture. This auction yard, from what I knew of it, practiced "good animal welfare," meeting the legal standards and norms of caring for animals during their time there.

Still, there were things I noticed that made it difficult to imagine the animals' well-being was not compromised. For instance, it was a hot summer day, and none of the birds had water as they waited for hours in the closed cardboard boxes. By the later end of the auction, the birds were panting when they were brought out of the boxes to show. This was not unique to this auction or the kinds of care it is mandated to provide. There is no legal requirement for birds to be provided water at any point in their raising, transportation, sale, or slaughtering, so their experience of thirst was a routine one—codified in animal welfare law (or, more specifically, farmed birds' exclusion from any US animal protection laws).[4] In fact, on commercial egg farms, induced (or forced) molting is a routine practice that involves withholding food and water to stress the bird's body so that when feeding and watering resume, the bird lays more high-quality eggs at a more prolific rate.[5]

I thought about the level of routine stress birds go through for egg and meat production. The high level of stress experienced by the birds in this chaotic and unfamiliar atmosphere was obvious. Hens were laying eggs in their boxes. One hen even laid an egg while being held up in the air on display. The audience erupted in laughter. Two hens escaped when their box was opened, and one tried to fly out into the audience but flew into one of the steel cables of the auction ring perimeter instead and fell to the ground. The stress of the animals heightened my own stress at watching this commodification of life unfolding. The space was a high-intensity limbo as the next phase of each animal's life was determined when the auctioneer called the winning bid.

Hundreds of animals sold over the course of three hours: hens, roosters, chicks of various ages, ducks and ducklings, geese and goslings, quails and quail chicks, game hens, rabbits, a single red golden pheasant (who elicited *ooohs* and *aaahs* from the audience and loud applause when they sold for a high price), and finally, three peacocks bundled and tied up, their legs bound, in plastic animal feed sacks. The auction also sold farm fresh eggs from chickens, ducks, geese, and quails at much lower prices than were available anywhere else. In addition to these nonfertilized fresh eggs, many dozens of *fertilized* eggs sold (including a set of four peacock eggs) to those buyers who wanted to incubate and hatch birds on their own at home. There were also other random items that showed up for sale: a rototiller, 16 boxes of cat litter, a bundle of hot wire fencing, rabbit and chicken feeders and waterers, and 110 forty-pound boxes of bananas.

That animals were being auctioned alongside bananas and cat litter only intensified the sense of their objectification and the way each animal's subjectivity was eclipsed or denied leading up to, during, and after the point of sale. Most of the animals sold for prices that I thought were incredibly cheap: $1 for a chick; $6 for a rabbit; $1.50 per dozen of fertile chicken eggs; $30 for a peacock. Some were higher value, such as the red golden pheasant, who sold for $105, or the six pedigreed mini lop bunnies, who sold for $40 each. But, in general, a little money goes a long way at the poultry auction. If you came to the auction with $20 to spend, you could go home with a five-week-old Toulouse gosling for $10, four young Silkie roosters at $1 each, and a dwarf Polish rabbit for $6. Or you could buy two boxes of cat litter at $4 each, a box of bananas for $6, and a bundled deal of two rabbit waterers and two feeders "that could use a good scrub" for $9. I'm not sure what I thought their prices *should* be or if there was even a price I would have felt was an adequate valuation for each of these animals' lives. Yet it wasn't actually the specific level of pricing that was disturbing but that these living beings could be priced and sold at all. It was that it almost didn't matter within the logic of commodification whether you bought someone's whole life, or a hunk of plastic fashioned into a feeder. Trying to consider and witness this stark reality while the event was unfolding in real time was overwhelming.

The practice of witnessing at an auction, and perhaps especially at the poultry auction, is an experience of complete sensory overload. It's organized chaos. The sounds of the animals alone—a cacophony of crowing, bawking, clucking, and squawking—are enough to drown out any coherent thoughts. The boxes on the shelves rattle and shake, and the sounds coming from inside them are almost made inaudible by the even louder cries of the birds being held up while the bidding unfolds. The low hum of the audience talking and laughing—and the occasional cheering—forms a background noise, and the auctioneer's calls permeate this auditory landscape.

The auction is also visually overwhelming. There are simply so many things to look at, and the animals in the ring demand so much attention in the urgency of their experience that it's difficult to notice other things. The art of noticing and attending the world around you requires care and deliberation, and this felt nearly impossible in this space as the frenetic energy took over. As I watched each sale and took notes on my phone, I was trying hard to listen to the details of the description of each animal from the auctioneer while keeping an eye on the animals themselves. *Wait, what breed was that chicken? Shoot, missed it.* Amazingly, it wasn't until an hour and a half into the auction that my gaze wandered slightly, and I noticed a giant screen just above the auctioneer's stand that reported the lot number of the sale and a short description of each animal. I had no idea how I had missed it. When I was sitting in the bleachers before the sale, I had looked around and read each sign on the wall around the screen but hadn't noticed the screen itself. I wondered as I was driving home that afternoon what else I had missed just by virtue of paying attention to one thing over another, and how I would notice different things (and things differently) the next time I attended a sale. I wondered what kinds of things I would never notice, being an outsider both to this community and to the day-to-day lived experiences and knowledge of farmers and auctioneers. And I wondered about the animals themselves and what and how *they* notice things that will, perhaps always, completely escape my attention.

Purchasing live animals at an auction or feedstore is not a familiar experience for most people. Unless you've been part of a rural community or have ventured out of the city to one of these places to purchase or observe the sale of animals, the auction can feel like an unusual sight. Certainly, my experience and fascination with auctions over the years are at least in part rooted in their novelty to me. I grew up in a city and lived in a city until my late thirties, and so I hadn't been enculturated into the specific practices of farmed animal commodity exchange that occur in rural communities. For me, and for others too, this type of commodification of animals might seem extraordinary or noteworthy. However, animals are commodities in all kinds of ways that may be more or less familiar depending on one's social and geographic location. Buying purebred animals as "pets," for instance, may be more familiar to some people. That it's possible to buy and sell a pedigree dog and cat highlights the ways that they can generate capital in the form of profit, and also that their pedigree adds another layer of cultural capital to their value, as I saw at the auction with the enthusiastic hum and applause of the audience when the pedigreed mini lop rabbits sold.

Critically attending to the circulation of living animals as commodities in spaces like the auction or the feedstore, though, helps to illuminate a much more familiar—more widely *everyday*—form of commodification that is woven into the fabric of rural, urban, and suburban landscapes: the commodification of dead animals in the form of meat, or reproductive "outputs" in the form of eggs and milk that are available at nearly every grocery store and restaurant across the country and in many places around the world. These relationships of commodification and consumption that are so thoroughly imbricated in human-animal relations can be difficult to attend to. I've found in myself over the years that this attending takes constant vigilance because there are powerful social and cultural forces that work to lull the mind into acts of forgetting and normalizing the nature of these commodity relationships.

Commodifying an animal like an egg-laying chicken is an abstracting process—the chicken's inner life, subjectivity, emotional connections, and kinship ties are obscured in the way she circulates as capital generated from what she can produce: eggs or meat. This is the logic underwriting the auction yard and the country feedstore Eric and I visited years earlier to purchase chicks. Even before they would produce eggs, those tiny chicks were already commodities, and our purchasing them was reaffirming and solidifying this status. We were entangled in a multibillion-dollar global trade in farmed animals, meat, dairy, and eggs.

To give a sense of the scale of the animal agriculture industry: In 2022, approximately 75 billion chickens, 1.49 billion pigs, 637 million sheep, 308 million cows, and 91 million tons of fish (who are counted by weight, not individual lives) were killed for food globally, in addition to billions of other animals, such as turkeys, ducks, geese, goats, and rabbits.[6] The scale at which the auction yard operates relies on local, relatively small-scale circulation of animal capital, but this local site of exchange is connected to these global networks of trade and statistics—materially and conceptually. This trade is fundamentally anchored by a reproductive and biological process, like the ovulatory cycle of chickens that enables egg production and consumption.

At the auction, the animals appear, arriving in vehicles, are registered, bid on, and sold, and then are paid for, picked up, driven away in vehicles, and disappeared. It can be hard to think about what it was that got them there in the first place—what conditions were necessary to transform them into a commodity. These are obscured through the act of commodification itself and the reproductive labor that shapes the animal into a commodity in the first place. To attend to this origin is to acknowledge that each animal's presence at auction, at the feedstore, and at the dinner table is the result of reproductive labor and control. It's impossible to fully comprehend the nature of the commodification of life without understanding this reproductive process and the way it affects the animals themselves.

Five years after we came home with that little cardboard box of chicks, we waited in the exam room as the veterinarian examined Emily. She had been looking droopy, and her abdomen felt unusually full and soft, so we were at the vet to have her checked out. The vet, dressed in scrubs printed with chickens of many different breeds, carefully felt Emily's body and listened to her heart and chest with a stethoscope. Slinging the stethoscope back around her neck, she said matter-of-factly, "Well, she's got quite a bit of fluid buildup in her abdomen. We'll drain that and it'll make her more comfortable. But I suspect that this is ovarian or some other reproductive cancer. There's no cure for that, so at this point we're just looking at comfort care."

I struggled to stop my chin from trembling and cleared my throat. "Ovarian cancer? Is that common? Is there anything we can do?"

"The only thing to do now is make her comfortable," the vet explained again gently, "and we can make the decision to euthanize her when she's no longer comfortable." She paused, allowing us to take this in, and then continued, "Ovarian and other reproductive cancers are really common in egg-laying hens."

As it turns out, it's estimated that as many as 30 to 35 percent of egg-laying hens will develop ovarian or other reproductive cancers by age two and a half.[7] What scientists refer to as "incessant ovulation" in chickens is shown to lead to these markedly high instances of cancers.[8] Chickens are one of the only animals other than humans who develop high rates of spontaneous ovarian cancer, which has led to their wide use as research subjects to better understand and potentially prevent human ovarian cancer.[9] However, the accuracy with which chicken cancer development can be translated to the human model of disease has also been the subject of doubt, and claims about the fruitfulness of preventing human ovarian cancers through studying chickens may be overstated.[10]

Prior to our living with chickens, I hadn't thought about any of this. I hadn't thought in any sustained way about the health impacts of this incessant ovulation on the chickens themselves. For five years, our flock had been healthy, and so while I knew from reading about chickens and talking to others who cared for them what could go wrong, or how their lives and deaths might be shaped by their breeding, I didn't know it by witnessing it. Prior to living with chickens, it didn't hit home for me just how extensively humans have shaped the chicken into such a prolific egg-producing bird. I hadn't thought about it in any targeted or sustained way; I thought that somehow there were just some species of birds with particularly spectacular egg-producing capabilities, and I knew that some breeds of chickens were more prolific than others. I didn't think about the obvious fact that, as in humans, eggs are the product of an ovulatory cycle. I wonder, even now, about the taken-for-granted assumptions that guide my understanding of the world that I don't even know are shaping what and how I think (or *don't* think, as it may be).

In *The Cow with Ear Tag #1389*, I wrote about how one of the common misconceptions about cows is that they just "make milk"—an idea that obscures the fact that mammals must have recently given birth in order to make milk. Intensive human intervention into the reproductive process is necessary in order for cows to be in a constant state of milk production. Cows are impregnated annually (typically via artificial insemination), and their calves are taken away shortly after birth so that their milk can be redirected for human consumption. Male calves are superfluous to the industry and are often slaughtered at a young age (usually after six months living in intensive confinement) for veal. Some males possessing what the industry refers to as "superior genetic heritage" are raised as bulls for semen production, where they are ejaculated for semen circulated as a global commodity to be used in artificial insemination. The lives of all animals in the dairy industry are shaped by the reproductive process—as breeders, as milkers, as sources of semen, as discards, and eventually when the reproductive process is exhausted, as meat. Without the reproductive process—of artificial insemination, pregnancy, and birth—there would be no milk. And without the continual impregnation of cows, and the diversion of milk meant for calves into the market, there would be no dairy industry.

Reproductive intervention in the case of chickens raised for egg production is somewhat differently organized—not only around artificial insemination/fertilization, pregnancy/incubation, and birth/hatching but also around the ovulatory cycle. Reproduction is done either by the extraction of semen from a rooster and artificial insemination of the hen, or fertilization of an egg through a direct encounter with a rooster. Female chicks enter an egg commodification cycle. Like the superfluousness of male calves in the dairy industry, male chicks have no use to the egg industry because of their inability to generate capital and are thus discarded as waste.

The reproductive programs central to egg and other animal-based food industries rely on the spatial and social segregation of the industry and, in most cases, on conditions of intensive confinement. Semen extraction occurs in one space, artificial insemination in another, incubation and hatching in another, and egg laying for human consumption in still another. In each of these segmentations, the birds' kinship relations are fractured—roosters are separated from hens, hens are separated from chicks and are denied the experience of parenting their young. Their capacity for prolific egg production to fulfill human appetites eclipses possibilities for their social, emotional, and embodied flourishing.

It is not happenstance or a natural part of evolution that has driven domestic chickens to lay so many eggs or dairy-breed cows to produce such high volumes of milk. In the case of milk, breeding practices have dramatically increased the volume of milk produced per cow: In the United States, over the past hundred years (between 1919 and 2020), the average cow's milk production has increased by seven times due to selective breeding (as well as because of the use of hormones and formulated feed).[11] A cow is milked two or three times per day (usually by machine) approximately three hundred days out of every year, leaving roughly sixty days as a "drying-off" period in the two months leading up to the birth of her calf. This cycle continues for several years until the cow is no longer getting pregnant as easily or producing milk so prolifically, at which point she is sent to slaughter. The reproductive history of domestic cows on which this cycle relies dates back to the domestication of a wild ox, called an aurochs, some 10,500 years ago in the Middle East and western Asia.

Similarly, chickens have been carefully selected for increasingly high rates of ovulation; most domestic chickens bred for eggs ovulate (produce an egg) roughly once per twenty-four-hour period, producing, on average, more than 300 eggs per year. In the 1930s, hens ovulated at roughly half that rate, laying around 150 or fewer eggs per year. It was in the early twentieth century that breeds were organized into "layers" (for egg production) and "broilers" (for meat production), after which their selective breeding split, following one path toward increased egg production and another path toward higher volumes of meat per bird. Attending to this breeding history, it's easy to see how humans have in recent history shaped the lives and reproductive capacities of these birds. However, selective breeding of chickens is not new; it has a much longer history, dating back to their domestication from wild red jungle fowl in 2000 BC and possibly even further—to 6000 BC—in Southeast Asia.

Like the impacts of intensive milk extraction on the cow's body, the exhaustive pressures of ovulation on the hen and the inadequate care she receives in most farming settings lead her to be "spent" after a couple years of egg production—production oriented around human consumption. *Consume: to do away with, to destroy, to use up, to spend.* Because dairy and egg industries rely so centrally on the reproductive process, when reproductive productivity declines, value is extracted in the only way that remains—the slaughter of the body for meat or the rendering of the too-worn-out-to-be-edible-by-humans remains. This is another routine reality that's often obscured by common conceptions of dairy and egg production as benign—the looming specter of slaughter and death that haunts every egg laid and every gallon of milk produced.

In those moments of learning about Emily's ovarian cancer, though, I wasn't focused on any of these broader realities of animal agriculture or on the selective breeding that led to her producing so many eggs. My attention was on Emily as she stood straight and still while the vet used a syringe to draw out the yellow fluid from her abdomen.

"We can do this one or two more times and it might give her some more good days, but we're on a downhill slope at this point, and you should watch her carefully to make sure she's comfortable. When she seems in pain again, bring her back in." As she spoke, the vet put together a bag of syringes filled with pain medication that we would inject into Emily's chest once a day to make her more comfortable. She showed me how to do the injection and then patted my arm and said before she left the room, "You guys are going to be OK. Just go ahead and check out at the front when you're ready."

After the vet exited, I gently picked Emily up off the exam table to put her back in the carrier, and I was surprised to feel how light she felt—just a wisp of a bird, golden feathers and bone.

As I drove to the auction that warm, sunny day in May years later, I took in the beauty of the landscape. Enumclaw sits in the flat land not far from the base of Mount Rainier in western Washington, to the east of the Cascade Mountains. Rainier looms large and majestic, white with snow, across the farm fields. The drive takes you through the Muckleshoot Reservation, and I thought about the history of settler colonialism and reservations in the United States and the close link between the auction yard I was visiting and this history. On the surface, the connection between a contemporary farmed animal auction yard in Washington State and the history of dispossession and violence involved in settler colonialism and capitalism might not seem readily apparent, but it's there if we attend carefully to it.

On the drive, I passed a fenced pasture that enclosed about a dozen bison—a mix of adults and calves. Some stood around, eating grass; others lounged lazily in the field. The calves stayed close to their mothers. With this book on my mind, I thought about how bison living today are an embodiment of haunting—their presence, in dramatically reduced numbers and in captivity, recalling the slaughter of bison in the 1800s that drove them to near extinction. Tasha Hubbard, a Cree educator and filmmaker from Peepeekisis First Nation in Treaty Four Territory in southern Saskatchewan, explains, "When Euro-Westerners began to eye the Great Plains as part of the imperialist project, they identified two major obstacles to claiming the land: Indigenous peoples and the buffalo."[12] Killing the bison was one central tactic that drove many members of Native Plains tribes to starvation through eliminating a primary food source and fracturing kinship and Indigenous nation-making relationships, as well as the spiritual and cultural practices oriented around these relationships.

Hubbard explains that Indigenous nations are constituted not only by humans (the way settler nations are) but also by a web of multispecies relations, and the bison were a core part of Great Plains peoples' nationhood.[13] The annihilation of the bison by settlers was an integral part of the elimination of these nations and the twinned project of constructing and solidifying the settler nation-state. But Hubbard goes a step further to attend not only to the effects of the slaughter of bison for Indigenous human peoples but also to the devastation for the bison themselves and their own intraspecies kinship relationships and intergenerational trauma as a result of what she describes as buffalo genocide.[14]

Bison grief is so profound that even when a herd was being actively hunted, their lives imminently at risk, they would stand around a fallen member of the herd and mourn, gathered as easy targets for their mass slaughter.[15] The intergenerational trauma of this killing and the related disruption of reproduction of the bison, whose lives rely on close social networks of support, drove the species further toward extinction.[16] Bison populations in North America were estimated to be between thirty and sixty million in the eighteenth and early nineteenth centuries; however, "by 1883 the buffalo were effectively removed from the Great Plains, relegated to remnant herds of orphan calves, a few animals in captivity, and a refugee herd in Yellowstone. Estimates put their numbers as low as a few hundred animals by 1889."[17] Meanwhile, cows and other farmed animal species proliferated as farms and ranches were established. Farmed animals were used to physically occupy the ancestral lands of Plains Indigenous peoples, bison, and other native species, as well as driving the destruction of the rich biodiversity of the Plains ecosystem itself.[18]

This process wasn't a simple story of removal and theft—transferring property from one owner to another and replacing it with another form of property. This solidification of ownership was one of first establishing propertied relationships with all of the lives and landscapes settlers would consume (in other words, first establishing that something and someone could be owned at all). This occurred through a variety of conceptual and material practices.

As Patrick Wolfe articulates, "Settler colonialism destroys to replace."[19] In the case of animal agriculture, Hubbard explains that it was necessary to "remove the existing species and replace it with one of European origins in order to solidify ownership of the land."[20] Animal agriculture offers a particularly salient illustration of this logic in that "it is inherently sedentary and, therefore, permanent. In contrast to extractive industries, which rely on what just happens to be there, agriculture is a rational means/end calculus that is geared to vouchsafing its own reproduction, generating capital that projects into a future where it repeats itself."[21]

The permanence enacted by agriculture's sedentary nature is cemented in part by the presence and reproductive regeneration of farmed animals on farms, in feedlots, at auctions, as they are transported on roads, and at the slaughterhouse. Farmed animals occupy space by the billions in the United States each year, their populations continually replaced through breeding programs. Agriculture's permanency is also solidified in the infrastructure and architecture, like fencing, needed to contain farmed animals and carve out more territory for farming. Fencing has a fraught history. It might be seen as an innocuous or neutral practical activity of enclosing land and animals, of demarcating space as property. It is that and more; as Reviel Netz describes in his book *Barbed Wire: An Ecology of Modernity*, fencing is a powerful technology of property ownership and control that has been integral to maintaining property relations, and the colonization and commodification of farmed animals in particular.[22]

As I was reading about fencing as a technology of settler colonial expansion, I thought about our own fenced-in backyard where the chickens lived. When we moved in, the fence was already there—aesthetically ugly cyclone fencing with bent links. We didn't like the look of it at all, but we still wanted the technology of enclosure that the fence provided, and we couldn't afford to replace it with a more aesthetically appealing fence. So, we planted vines and bushes along the fence that would, over the years, grow so thick and green that they obscured the whole fence, a beautiful green wall that bloomed at different times during the year. I realized reading about the history of fencing that my desire to not look at the fence but still benefit from it was perhaps a core feature of how settler colonialism is sustained—I didn't like to look at its ugliness, but I enjoy its effects.

The Goenpul Indigenous studies scholar Aileen Moreton-Robinson argues in her book *The White Possessive* that possessive logics that rely on dispossession normalize the everyday workings of settler colonial society. As I read her book, I thought about how, even in something so seemingly innocuous as raising backyard chickens for eggs, it was dispossession that enabled my possession of this space and these animals—dispossession of those who are still here in the place that I live and those who are gone, eliminated, there only through haunting reminders of their presence. And I also thought, as I was sitting at the poultry auction, about how auctions are a market-oriented mechanism of possessiveness through dispossession—dispossession of the lives that circulate in the auction ring through the changing hands of ownership (possession).

Consumption—of animals at auction or of eggs at home—is a process of incorporating something into one's own circle of control. It is a process of absorption, of "making mine"—to own, to subsume, to do with it what one will. The consumption of animals dispossesses them of their lives and selves, and attending to this highlights the irreconcilable harm done by crafting and solidifying notions of property, and life as property in particular. These are harms that deepen and intensify across long spans of history into the present, and they are concepts that haunt the very bedrock of society today.

When we returned home from the vet with Emily, she was relieved to see her flock—they preened each other and clucked softly. She perked up and was back to her regular life routine for a while. She and the rest of the flock had spent their days over the previous five years roaming freely in our fenced-in backyard and sleeping safely in their coop at night. They had favorite spots for dust bathing, carving out dusty little ditches in the yard where they would lie down and throw dirt over themselves until it penetrated their feathers and reached their skin. (Dust baths are essential for keeping chickens free of parasites, like mites and lice, and so they need ample room to engage in these cleaning rituals—something that birds in intensive confinement do not have.) The flock hunted around the yard for bugs and other things to eat to supplement their diet of a complete feed we provided them. Year-round we also gave them treats from the kitchen: watermelon, berries, cabbage, chard, kale, lettuces, and oatmeal on cold winter days. They helped themselves to everything we were growing in the garden (raspberries, strawberries, leafy greens, turnip and radish tops, broccoli, etc.). Exploring the yard, they found interest and variety—curiously inspecting any new feature that appeared, like a new wheelbarrow or shovel, and assessing how it fit (or didn't fit) into their lives.

Our care of the chickens evolved as we learned more about them and about the impacts of egg laying on hens. Several years before Emily got sick, at the recommendation of some folks in the farmed animal sanctuary community, we had started hard-boiling and crushing eggs, then feeding them (including shells) to the chickens. This was a way to replenish the nutrients lost in egg laying. This didn't, of course, stop the egg laying, and so didn't reduce the chance of ovarian cancer, but it did help them nutritionally, and they all lived long lives for egg-laying chickens: Emily died at five years old, Charlotte at seven years old, and Jane and George both died in their eighth year.[23] We saw this practice of feeding the eggs back to the chickens as a small act of care and making amends, an attempt at repair, however small, for our having taken their eggs to begin with and for the more insidious impacts that humans like ourselves had wrought on chickens' bodies through breeding and consumption. They devoured their eggs hungrily, pecking up each last tiny bit of shell, white, and yolk from the dish we put out for them each day.

Living with the flock prompted regular reflection on the relationship I was in with each of these hens. This relationship shifted first when I cooked that dead chicken who catalyzed Eric's and my path to eliminating animals from our diets. It changed again when I learned that we could feed their eggs back to them and I realized that just because *they* were laying eggs didn't mean *I* was entitled to eat them. This evolving relationship made me consider the kinds of relationships I had with other farmed animals. When we initially decided to raise chickens for eggs, we were enthusiastic about being in relationship, for the first time in our lives, with the animals who were the source of our food. But this wasn't the first or only relationship we had with farmed animals. In consuming and commodifying animals, we had always been in relationship with them—relationships oriented around how they (abstracted others raised in production spaces) and their bodies could fulfill our desires to eat meat, dairy, and eggs. I realized later that in bringing these chickens into our lives, we were curating a relationship of closeness and care that could make us feel better about the other less intimate, more violent relationships we were in with farmed others. Our early relationship with these four chickens, then, despite the ways that we cared attentively for and about them, was still one of instrumentalization and extraction—instrumentalizing their reproductive capacity and the long history of breeding for egg production, and extractive as it stole from them their life and reproductive energies. We instrumentalized their lives and the way they fit into ideas of "humane farming" not only to justify our consumption of their eggs in those early days but also to assuage what discomfort we felt about the animals we ate—whose lives and life energies we extracted—who were *not* in our care.

Our consumption of animals made it difficult to even be aware of, let alone undo, these types of relationships. Our possessiveness (through occupying, owning, and eating) and the entitlement we felt made it an impossible task to imagine worlds otherwise. Outside of the overwhelmingly dominant relationships of consumption and commodification in which many humans and farmed animals are entangled, how might we begin to imagine other futures? What other forms of relationality are possible? And how can these be wholly oriented around the experiences, well-being, and flourishing of those others who are so readily commodified and consumed? What alternative forms of care are possible that aren't oriented in and through our benefiting in concrete ways from the consumption of others?[24]

As we navigated life with the chickens, we tried to take these questions into consideration and find ways to live more gently and attentively with these individuals. In addition to being attentive to the kinds of treats they enjoyed, the way they liked to navigate the backyard, and the kinds of interactions they liked to have, we considered these questions at length in their end-of-life care and deaths. When each member of the flock died, we buried them in the backyard, creating a ritual where the other hens were able to observe their dead kin to understand what happened, and where we could all say goodbye to this individual we had loved. Many keepers of backyard chickens, we learned, will either slaughter and eat the chickens themselves or pass them on to others who want to slaughter and eat them. Creating a ritual around burial, for us, was another way to push back against the norms that frame chickens as food and commodities. Even their veterinary care disrupted notions of who chickens are, what they're owed in their flourishing and care, and how their species is categorized. We were in a position to seek out a bird and exotic pet clinic in Seattle, which, at the time, was the only facility to provide comprehensive veterinary care for chickens. Both the rarity of these veterinarians and the costs associated with accessing this care make it difficult to transform our relationships with farmed animal species in even the most mundane and practical ways. A reconfiguration of how we relate to animals who are routinely farmed is, then, not just a conceptual project of undoing common ideas about these species; it's also a practical and logistical project that involves the mundane aspects of everyday life—from daily care to medical treatment to end-of-life care and dying.

A few years after buying the chickens, Eric and I began volunteering at Pigs Peace Sanctuary in Stanwood, Washington. Farmed animal sanctuaries are spaces dedicated to the rehabilitation, care, and creation of livable lives for those animals who have been routed out of a commodity circuit or site of exploitation. Pigs Peace provides a permanent home to potbellied and big farm pigs (at the moment of writing this chapter, there are 101 pig residents at the sanctuary) who have been rescued from agricultural and other situations of abuse, neglect, and abandonment. The sanctuary space, relationships, and systems of care are oriented wholly around the flourishing of the animal residents who live there. Founder Judy Woods, staff, and volunteers labor in service to the pigs, dedicating their time and energies to crafting a radically different world for farmed animal species in the microcosmic universe of the sanctuary grounds. Pigs Peace imagines and manifests these different futures through the mundane practices of the everyday that make up the lived experiences of the residents. The world-crafting work unfolding at Pigs Peace is also the work of making amends to the individual pigs who live there, helping them to heal from the haunting pasts from which they were rescued. It's also a way to make amends for these much longer, more far-reaching histories and contemporary practices of exploiting animals for food. Pigs Peace, then, is as much a place of undoing as it is one of making.

Uncovering the harms that agriculture delivers on farmed animals and, importantly, recognizing them *as harms* involves an acknowledgment of how we may be implicated in harming others and what kinds of transformative action this might require. But a focus on harm alone may not be powerful enough to transform these relationships. Witnessing what's going on at Pigs Peace helped me to understand this. Certainly, the sanctuary can be a space to witness the past impoverished lives animals lived in production spaces as a result of being subjects of farming, reproductive control, commodification, and consumption. These are the life histories that accompany the animals who arrive at the sanctuary, they are the experiences from which the animals may need to heal, and these realities are the reason the sanctuary exists—to offer an alternative and an opportunity (hopefully fulfilled) for flourishing to the extent that that's possible for each individual.

It's this possibility of flourishing that can make it possible to better understand harm. Many people probably know to some extent if they think about it that farmed animals experience harm in food production. When we purchased those four chicks, I knew in a general way that commercial production spaces involve suffering, confinement, bodily control, and premature death. This was one of the primary reasons we wanted to raise chickens in the first place: to eat eggs that we knew were locally sourced from chickens whose lives we knew and shared. When I began to get to know the chickens and what mattered to them, I began to understand more clearly what's routinely denied to birds who are raised for food. And it wasn't until I visited a sanctuary for the first time that I came to understand that alternative conceptualizations of farmed animal species and the lives they live could exist. Intuitively, for instance, we might know that gestation crates or battery cages are spaces of misery for pigs and chickens, but until we see the reality of flourishing at a space like a sanctuary where they are not farmed, it's hard to fully understand and internalize the violence done to them in farming. It's difficult to see what's lost when you don't know what they could have had or what their flourishing could look like. It is this flourishing and the tragedy of its absence that haunt the space of the farm. And it is the memory of this absence that haunts the sanctuary.

These hauntings are sustained by the normalization of consumption and commodification. In sharing my story of how purchasing and raising chickens for eggs changed me, my thinking, and my relationships with farmed animals, I've been told by numerous people involved in rescue from backyard chicken situations that this is unusual. They explain that backyard chickens don't typically create a world-shifting rupture but instead fit neatly into reaffirming the acceptability of farming animals in the first place through personalizing it. I'm honestly not sure why that rupture occurred when and how it did. Eric and I had formerly been firmly in the camp of those who believed that eating products from animals raised in alternative conditions to factory farms was an important form of "voting with our dollar" (or a capitalist-oriented activism, if that wasn't an oxymoron). It made sense to us that rather than opt out of animal consumption entirely, we would support farmers who were trying to offer a small-scale alternative. We saw these as two polar opposite practices—industrial versus small-scale production. But there was something for us about going through this process of raising and living with chickens, paired with my growing education as an animal studies scholar, that illuminated an inkling of awareness of the logics of commodification, disposability, possessiveness, and property relations that underwrite farming and consumption practices at any scale.

As I sat at the poultry auction, I had a strange experience that I hadn't noticed at previous auctions. When I attend an auction, I enter the space resolved not to buy any of the animals. I am there to witness, to be present for what is unfolding, and then to share this in my work as a way to hopefully illuminate and change the way animals are thought about and treated. My intention is not to financially contribute to this system, although this has been the cause for ethical pause many times. I don't want to support and reaffirm the conceptualization of the animal as a commodity by using the currency that matters most to this system of commodification. However, my principles about not supporting the commodification of the animal through purchasing them would hardly matter to the turkey who might have a long and fulfilling life at a sanctuary if I had purchased her for twenty-two dollars. And so, this internal torment was a familiar experience sitting in the auction yard. It felt normal, and I could almost visualize the well-worn grooves in my brain that moved me through this circular thought process, rarely coming up with any new insights, always ending in a frozen state of inaction. It also felt familiar that over the course of one, two, three hours of watching chicken after duck after rabbit after goose sell, the resolve to not buy any of the animals softened, wearing down in the face of such routine objectification, until I felt a low-grade desperation to buy even just one to set them on the path to a different future.

But what was new and felt different this time at the poultry auction was the way my thinking also settled at certain moments into a different set of grooves—familiar not in the context of the auction but in the context of my life as a consumer more generally. In this site of frenzied consumption through competitive bidding, I felt a flicker of excitement, an almost automatic pull to consume, to jump into the fray, to bid and buy. I could see the excitement and glimmer of satisfaction in the eyes of the winning bidders in the audience when the auctioneer called the sale in their favor.

I identified with that momentary feeling of pleasure in buying something that I desired. As quickly as it came, though, that flicker of excitement I felt was gone, and what was left was a hen held up in front of the audience by her wings, eyes wide, panting, her feet kicking at the air—our final glimpse before she was stuffed back in a cardboard box and reshelved.

There is perhaps something more that needs to be excavated and transformed—not just specific commodity relations with animals and other forms of life, but also the persistent orientation to the world itself organized through these dynamics of commodification (what we can buy), property (what we can own), and consumption (what we can devour through destruction). Perhaps working to undo this deeper logic can change the kinds of relationships that are possible.

A few weeks after our visit to the vet, Emily died in my hands as I held her—the hands that had scooped her up as a tiny chick five years before and purchased her at that country feedstore. *How much had changed since then. How much you have taught me*, I thought, and kissed the top of her head. I laid Emily gently down on the grass, and her flock gathered around, pecking her body and then waiting for her to wake up. When she didn't wake up, they stood in a half circle around her remains. We were gathered together in unusual silence to mourn Emily and to acknowledge her life and our loss.

Emily, George, Jane, and Charlotte are all now long dead, but I reflect often on their lives and on my own life lived in relation with them. They changed what it was possible for me to know and acknowledge about the effects of my own desires. It was through my relationship with them that I came to understand the need to revoke my sense of entitlement to buying and consuming someone else's body and life as if it was mine to own, to destroy, to absorb, or even to consume within a narrative of care. They changed how I thought about and practiced care oriented around how to support the lives and flourishing of others in the ways that mattered to them.

How we come to know and experience the world shapes how we can be in relation to others. And how we are in relation to others shapes how we come to know and experience the world. This is always unfinished work that requires vigilance and openness to being changed and to learning how to live more carefully and gently so that we are haunted not by the pain we have caused others through the fulfillment of our own fleeting desires but by the beauty of the imperfect efforts we have made to change the very nature of those desires.

5

Ghosts in the Garden

What does it mean to think about those histories that are
difficult to remember well—either because the present in
some way requires erasing what happened in the past or
because particular past events have become so taken-for-
granted that it is hard to imagine that the world was once
different, and how?
—Alexis Shotwell, *Against Purity*

I stood across the street and watched a bulldozer barreling along, indiscriminately plowing down trees, bushes, and underbrush to clear the overgrown city lot around the corner from our house in Seattle's North Beacon
Hill neighborhood. As long as we had lived in the neighborhood (ten years
at that point), and clearly much longer, this lot had been a densely wooded
haven for all kinds of critters—over the years, I had watched crows, sparrows, Steller's jays, and juncos nest in the trees. Several generations of raccoons had raised their young, and I regularly saw rabbits pop their heads
out of the low thicket of blackberries that covered the ground. Sometimes
on our nighttime dog walks we would see a raccoon lumber noisily out of
the brush and down the street or a rat scurrying along the sidewalk and
then diving into the underbrush when they saw us. One year, I watched

squirrels build and tend to their nest, busily scurrying up and down the trunk of the tree and jumping from branch to branch. Another year, crows nested in one of the tall trees, and I watched with joy as their two loud fledgling crows learned to survive in the world. A familiar troop of four, sometimes six, flickers who lived in the neighborhood could be found peck-peck-pecking at a dead tree trunk, stopping periodically to eat the insects they had uncovered. I had picked up many snails over the years on my early morning walks and moved them carefully in the direction they were headed, back to the edges of the lot so that they would not be crushed by the morning human foot traffic, after a night out. And when the rains washed earthworms out onto the sidewalk, I would carefully pick them up and place them back into the rich loam and decomposing leaves that spilled out of the lot onto the sidewalk, out of harm's way.

I used to think when I was out walking the dogs that this little wildlife oasis in the neighborhood couldn't be around much longer; the neighborhood's real estate market was booming, there was a housing shortage in Seattle, and developers were approaching people throughout the neighborhood to purchase and demolish their houses for modern new-construction single-family homes and townhouses. "There's no way this lot is going to last," I'd said aloud to the beagles just the week before. I didn't know who owned the lot or if it had ever had a structure on it at all. But I liked to imagine that this little patch of land had somehow always been home to generations of urban wildlife (and forest wildlife before the place was urban) and that it would be preserved long into the future as their home.

This daydream was dashed jarringly when I watched the bulldozer crash through the wooded lot with no warning, no advance preparation or clearing of trees. I didn't know you could just bulldoze an entire lot, trees and all, until I had been working on my computer at the kitchen table several years before and suddenly a bulldozer plowed into and knocked down the old flowering cherry tree in the lot next door, just outside the window. It was a loud, sickening crack as the trunk split and a great crashing sound of branches and leaves being crushed against the earth. I had watched a pair of hummingbirds nest and raise their young in that tree earlier that spring. The neighbor's was a double lot owned by a family who had been in the neighborhood for generations. There had been a small modest home on one of the lots, and the adjacent one was maintained as their yard—a parklike setting and home to its own community of urban animal residents. One of the adult sons in the family had decided to build a house spanning both lots, and so the original structure and lot next door were bulldozed to bare earth; then, over the following two years, a large three-story house occupying most of the two lots was constructed.

Seattle's Beacon Hill is a rapidly gentrifying neighborhood southeast of downtown that stretches north and south along a ridge overlooking the industrial district. With this process of gentrification has come an influx of white, middle-class residents (like us), a dramatic increase in median income and median housing prices, the displacement of long-term human and nonhuman residents, and the massive development of new-construction condos, townhomes, and single-family homes. As part of this process, gentrification in Beacon Hill has also involved an attention to environmental aesthetics. For instance, Jefferson Park, a 110-year-old park in the middle of Beacon Hill, is composed of a community center, tennis courts, basketball courts, sports fields, picnic pavilions, walking paths, a skatepark, a spray park, a state-of-the-art playground, and a public urban agriculture project called the Beacon Food Forest. This park and its many amenities are frequently used as selling points for the neighborhood: rental and for-sale listings often mention the property's proximity to the park, and city funds were invested in a major renovation to make the park even more desirable for the changing population.

As spaces like the park were transformed to accommodate the rapid course of gentrification, I also noticed the elimination of the "wrong kind" of green spaces, like the abandoned lot I watched bulldozed to the ground, or trees that are in the wrong place for optimizing people's views, both of which contribute to reducing overall space in the city available to urban wildlife for habitat and nesting. These different kinds of creative and destructive ecologies are primarily organized around, and driven by, humans' use of space, often with little regard for the way other species use and inhabit these environments, too. Even as the "abandoned" lot was certainly not "abandoned" by the many animals who made their home there, it was of no use to human residents of the neighborhood in its existing state—in fact, it wasn't even possible for humans to enter the lot without hacking away at the dense wall of blackberry bushes that protected its boundaries. Overgrown and "unused" (read: undeveloped) lots are often associated with the idea of "blighted" neighborhoods, seen as drawing down home prices and as promising sites for developers to make a profit. There was no way—between the changing middle-class sensibilities of the neighborhood and the opportunistic development unfolding—that the wild and unkempt corner lot, home to so many other species, would survive.

As the bulldozer began its work, clearing the way for what would become a modern three-thousand-square-foot, single-family home, the animals fled. Crows, flickers, sparrows, squirrels, raccoons, rabbits, rats, a Cooper's hawk, even a swarm of wasps scattered frantically, given no warning of their sudden exile. It was unusual to see the raccoons and rats during the day, but their slumber was disrupted by the destruction of their home, and they had no choice but to tumble, bleary-eyed, from their sleeping places. I stepped to the side, out of the path of the fleeing animals, astonished at the sight. I had never seen anything like it. I worried over the smaller or slower beings who did not get out—the snails, the slugs, the spiders, the butterfly and moth cocoons, the older or slower mammals, and who else? Who would be ground into the soil by the bulldozer's great tracks? And I wondered, for those who survived, where would they all go? Where would they make their homes, raise their young, store their food for winter? Where would they find spaces to be invisible to the ever-encroaching incursion of humans into their lives and lifeways?

The question of *home*—whose home, who belongs, and who doesn't belong in a place—emerges at the heart of the process and politics of gentrification. Displacement is not just about physical displacement or alienation from a geographic place identified as *home*. The creation of home is made in and through the relational sense of community and cultural practices that can be ruptured when these projects of development occur. This is the case with humans and also with animals. The *place* in *displacement* is as much an emotional and relational location as it is a physical or geographic one. Places, as geographies and relationships, make us who we are, and to be dis-placed (*dis* is a Latin prefix for *apart* and *away*) is to be rendered separate or alienated from those life-sustaining roots. Displacement does not occur spontaneously without a cause; it is the result of someone or something *doing* the displacing—a pioneer family felling trees to build a log cabin on the Duwamish Tribe's and Pacific Northwest native animals' land, a middle-class white person moving into a working-class community of color, a change in city housing policy, a bulldozer clearing a city lot. Displacement is not an innocent process. It does irreconcilable harm to relationships to place, social connections, kinship networks, and even the very possibilities for survival.

It felt like all around us the habitats for urban animals, and these urban ecologies themselves, were being destroyed with accelerating efficiency. Just a few days earlier, the owners of a new-construction home just a couple blocks away had cut down an eighty-year-old cedar and a large madrone tree—both native to the Pacific Northwest—that were blocking their view. The house had been built earlier that year, and the builders had carefully preserved the two trees, going to extraordinary and unusual lengths to ensure they were not damaged in the construction, and designing and building the house to accommodate their towering presence. The two trees had grown side by side, creating a little micro-canopy in a neighborhood struggling to retain its large-tree canopy as development proliferated.

When the madrone was cut down, the arborists discovered that one of the branches had grown around the city power line, fully encapsulating the wire so that it was impossible to remove the branch without cutting the line. They carefully cut on either side of the segment wrapped around the wire, leaving a two-foot piece of the branch hanging behind. Now, years later, that small length of branch remains, a haunting reminder of the tree that once stood tall in that place—merely a curious oddity to those who didn't know what had been there before. It hangs there like a ghost—once living and no longer part of its living whole, and yet persisting in a refusal of its erasure. In *Arts of Living on a Damaged Planet*, Elaine Gan and her coauthors write, "As humans reshape the landscape, we forget what was there before. . . . Our newly shaped and ruined landscapes become the new reality. Admiring one landscape and its biological entanglements often entails forgetting many others. Forgetting in itself, remakes landscapes. . . . Yet ghosts remind us. Ghosts point to our forgetting, showing us how living landscapes are imbued with earlier tracks and traces."[1] Someday that power line will need to be replaced and with its removal will disappear the last visible trace that that tree was there at all.

What are these earlier tracks and traces, and who are these ghosts that linger to work against the erasure of forgetting? Gentrification, like settler colonialism, is a process of attempted erasure—of identifying for appropriation, assimilation, and annihilation that which was here before. Although gentrification and settler colonialism are distinct phenomena and processes, they also have important things in common: Most relevant for the story I'm sharing here is the settler's entitlement to space at any cost and the way this entitlement shapes notions of belonging and unbelonging. Eve Tuck and K. Wayne Yang explain, "Settler colonialism is different from other forms of colonialism in that settlers come with the intention of making a new home on the land, a homemaking that insists on settler sovereignty over all things in their new domain."[2] Gentrification involves a similar process of "homemaking" through occupation—a claim to belonging that normalizes and naturalizes the gentrifier/settler (often two subjects encapsulated in the same body).

I want to take a moment to briefly track, in broad brushstrokes, back to early settler histories in Seattle both to provide context for contemporary gentrification in the city and to draw connections in shifting relationships to the land and other species shaped by whiteness and middle-class virtues. Both gentrification and settler colonialism are processes that dramatically affect humans—in particular, Indigenous communities, people of color, and communities with low income—and there are large and robust literatures on these topics. My aim in this chapter is to think through the impacts on other species, not with the intention to dismiss or overlook or sidestep the effects of these processes on humans but instead to draw attention to a less frequently acknowledged form of violence occurring alongside and entangled with their effects on humans.

All settler cities are built on histories of destruction, dispossession, and environmental ruin, and Seattle is no exception. To understand the shifting place of wild and domesticated animals in Seattle's history, I used Frederick L. Brown's *The City Is More Than Human: An Animal History of Seattle* as a guide. As Brown explains, when settlers arrived, the Puget Sound region was home to diverse Coast Salish peoples, rich forest ecosystems, and a wide range of wild animal species, such as wolves, lynx, bears, elk, deer, beavers, raccoons, minks, martens, wood rats, muskrats, and beavers.[3] In the early 1800s, the fur trade was an important industry and source of wealth for settlers and Indigenous communities in the region. The skins and furs of these animals—especially beaver furs—were appealing to European settlers as sources of capital, and Coast Salish hunters also traded these furs and skins for other goods that Europeans imported from other parts of the world through the Hudson's Bay Company.[4]

The fur trade as a major industry, however, was fairly short-lived. As it declined, animal agriculture grew, and with it the number of cows, horses, pigs, goats, chickens, sheep, dogs, and cats multiplied on the landscape in the mid-1800s.[5] Just as farmed animal species had been used to colonize the East Coast and the Great Plains, they were also used to seize land from Coast Salish peoples in the Puget Sound region, dramatically transforming the landscape and cementing ideas about domestication as civilization, a colonization of nature. This process was, in part, one of normalizing and naturalizing European norms and desires and their physical occupation of space. But it also was and continues to be enabled by an underlying sense of entitlement to space—entitlement as settlers and as humans in a rich multispecies landscape.

For wild animals around Elliott Bay, the arrival of settlers in the area marked a massive destruction of their habitat and food sources, as forests were felled for lumber and land was transformed into industrial sectors, residential housing, and pasture for farmed animals. This process of habitat destruction has continued over the past nearly two hundred years, culminating in the bulldozer I watched wiping out one of the few tiny patches of habitat remaining. That single bulldozer did not appear on that single lot in a vacuum. Its presence and the work it did were connected to these long histories and to the complex processes of environmental ruination. What would it have been like to be a bear or an elk or a muskrat in the region prior to settlers' arrival? What would it have been like to watch your dense forest home cut down, your kin killed or displaced, and your very connection to place ruptured? What would it have been like then, and what would it be like now, in the twenty-first century, as you hang on to that last tiny scrap of land before it, too, is destroyed?

Certainly, there were many animals who were actively hunted and killed in those early days (and this hunting had significant effects on the survival of some species), but one of the primary mechanisms through which a wholesale displacement of other species occurred was the transformation of the environment—the spaces and places that other animals relied on for their very survival. *Displacement*: a severing.

As this project of landscape development escalated, the topography of Seattle transformed unrecognizably from its "settlement" in 1850 through the end of the nineteenth century and into the twentieth century. Seattle is still a hilly city, but as settlers sought to develop the area, the steep hills posed a geographic problem that led to a series of regrading projects to make the land more amenable to the development of residential housing and industry. Former Governor Eugene Semple developed a plan for cutting a canal through Beacon Hill from the Duwamish River to Lake Washington, a project that began in 1901 but soon failed as a result of repeated cave-ins and increasing costs. The earth that was moved during this project, along with sediment dredged from the Duwamish River, was relocated to fill in the Duwamish tideflats to create the flat stretch of land at the base of Beacon Hill needed to build a large industrial district adjacent to the growing port in Elliott Bay.

To further develop this area as an ideal location for an international port, between 1913 and 1920, the curvy and meandering Duwamish was reshaped into a straight waterway designed to accommodate large shipping vessels.[6] This was a reengineering of the river with devastating effects both for the riverine and estuary ecosystem and for the Indigenous communities who relied on the Duwamish for food, such as salmon, herring, clams and other shellfish, and water birds who inhabited this ecosystem, as well as for those animals themselves. Coll Thrush, author of *Native Seattle: Histories from the Crossing-Over Place*, explains that "the creators of Seattle's new urban ecology thought they were improving nature. [Seattle city engineer Reginald H.] Thomson, for example, called the Duwamish's natural curves 'ugly' and 'unsightly,' preferring a channelized and useful river to one that was messy and unpredictable."[7] Today, the region is haunted by this kind of thinking—of taming, controlling, and refining "nature," and then naturalizing human development projects and their impacts on the landscape. It was a dominant way of thinking then, as it is now, with the prolific spread of new construction and the destruction of "unsightly" urban natures, like the bulldozed lot. By contrast, for the Duwamish, the river was a vibrant source of life, food, and cultural meaning, its curves and natural flow elements of beauty and flourishing. It was part of an interdependent landscape that necessitated respect, care, and reciprocity to maintain balance so that the river and all that's a part of it could be sustained healthily far into the future. Today, the Duwamish River Community Coalition works to heal the river and empower the still not federally recognized Duwamish Tribe to shape the restoration efforts in ways that are most meaningful to them.

Historically, as Seattle's landscapes themselves were reshaped according to beliefs about modernity and civilization, so too were norms about which species belonged. Farmed animals were species whose status as belonging/not belonging in Seattle shifted according to changing middle-class sensibilities. Initially, these species were seen as markers of civilization because of their domestication. As Brown explains, farmed animals became markers of wealth and prosperity in the city, luxurious sources of milk, eggs, and meat, and central to crafting settler ways of life in the region. But as urban development proliferated, driven by a preoccupation with colonial understandings of modernity, keeping farmed animals in the city became a symbol of "backwardness," and laws created restrictions on where farmed animals could live and graze, such as the one passed in 1907 that banned cows from grazing in the commons, or racial restrictive covenants that barred people of color from living in certain areas and simultaneously banned chickens and other farmed animals from these white, affluent neighborhoods in the 1920s.[8] In 1957, chickens were effectively banned from the city as a whole with the introduction of new zoning laws limiting the number of animals allowable on a standard city lot.[9] This has shifted yet again in recent decades, as chickens have become *en vogue* with white, middle-class Seattleites, like myself, and laws have changed again in response, allowing for keeping hens (but not roosters) on a standard city lot.

The categorization of nonhuman life was an important part of defining belonging and not belonging in the city. The distinction between *domesticated* and *wild* reflected ideas about who was *civilized* and *uncivilized*. Within the category of *wild* animals there were those who were seen as *resources* and those who were framed as *pests*—sometimes overlapping categories, as in the case of predator species who could be both resource (for fur) and pest (a threat to farmed animals). Within the category of *domesticated* animals, there were *livestock* and *pets*, again sometimes overlapping categories. But, as Brown writes, "As working animals left the city, the categories of livestock and pet became ever more divergent. To remain in the city, animals had to become pets. The other choice was to become livestock and reside in the country."[10]

In today's Seattle and elsewhere, these categories remain central to the way nonhuman life is conceptualized and treated. There is still considerable overlap in some areas—for instance, chickens in the city would be defined by some chicken keepers as both livestock (an animal used for food) and pet (an animal kept for affection and affective labor). In the case of urban wildlife, there is similar blurring for some species: some find squirrels and crows to be fascinating urban animals, a welcome addition to the urban landscape; others find these species to be a nuisance. In other cases, there are some persistently clear lines between those who are desirable free-living urban animals and those who are undesirable: wild urban rats, for instance, are reliably categorized as pests, marked for eradication. The alignment of these categories with notions of belonging/unbelonging is not merely an interesting exercise in classification; it has life-and-death consequences for those who are being categorized—consequences that reproduce themselves across generations—decades, centuries, even millennia. There are many species for whom this is the case, but I'd like to think, in this chapter, about rats as a long-maligned urban animal because they represent such a clear and consistent example of an animal *not belonging* and the unchecked hostility toward them that is born out of a sense of human exceptionalism.

I was struck when reading *The City Is More Than Human* that in our decision, as a white, middle-class couple, to keep chickens for eggs in the backyard we had unwittingly reproduced the incursion of farmed animal species, domesticated and "civilized," into an area already inhabited by a range of urban wildlife. Here we were, implicated in contributing to the violence of gentrification and settler colonialism, bringing chickens as domesticated co-occupiers of this patch of land. I wonder, now, at all the urban wild animals who would have been displaced or who no longer found the yard a safe and desirable place as a result. This experience offers a way to think through some of the dynamics that unfold in what is ultimately rooted in a human (and, often, settler human) entitlement to space.

This is a story first about the flourishing of rats and then their attempted annihilation for simply being a species that, according to human norms around animals' respectability, does not belong in human spaces. Although many humans (white or not, middle class or not) despise rats and believe they do not belong in places where humans live, my entanglements with rats and the white, middle-class next-door neighbors felt very much like a clear example of the mechanisms of gentrification and, beneath that, settler colonialism, as well as the specific effects of these intertwined processes on urban animals.

A few years into living with chickens, I started seeing rats out in the backyard in the daytime. The feed offered to chickens in urban areas often attracts rats and other rodents to chicken coops and surrounding areas where feed might have spilled. This was the case in the yard. At first, I saw just one or two rats every now and then; later, presumably as they realized the backyard was relatively free from predators, they became bolder—spending longer hours scavenging the yard for food, and in greater numbers. I started recognizing certain rats as they made the backyard their home. There were big rats and small rats, old ones and young ones, some who were grayer and some who were a solid brown, and there was one with a crooked tail who was always easy to spot and whom I developed a particular affinity for. I observed that, with the whole backyard to roam, the chickens and rats kept to themselves, clearly noticing each other but keeping a safe distance.

There is a common misconception that seeing rats out during the day is a sign of a booming population nearby, or that rats out in the daytime are sick. In fact, rats travel around and feed at all hours. They sleep in short intervals and forage for food at times and in conditions where they feel safest. Daytime is usually riskier because their visibility makes them vulnerable to humans and predators who would harm them, and so, especially in high-traffic urban areas, they are less likely to travel into the open in daylight. However, if rats learn that a particular area is relatively safe during the day, they will take advantage of the opportunity to look for food. Rats have a hierarchical social ordering within colonies, and rats out in the daytime may also be those who are lower in the hierarchy and thus have to forage at less opportune times than those who are more dominant in the colony.[11]

This was information I learned shortly after I began seeing rats during the day. I remember the first time I saw more than one rat at once in the backyard, I gasped, and felt a flutter of fear in my stomach. My conditioning, shaped by norms about rats as "vermin" to be reviled and exterminated, came so naturally as to evoke an almost automatic, visceral, and embodied response. Alarmed by the situation, I did some research and then talked to our Cambodian neighbors, who were long-term, cross-generational residents of the neighborhood.

"Yep, there are just a lot of rats on Beacon Hill—we may not like it, but it's just the way it is," one of the neighbors told me, which was a sentiment echoed by others. Nonchalant, another neighbor explained that they sometimes "put out traps or bait stations, but it doesn't do much, and it definitely doesn't get rid of the populations." In a city undergoing major development and new construction, rats are disturbed from their homes and so move to the neighboring lots—a process that increases their visibility. In my conversations with the neighbors, the consensus was a general acceptance that rats were part of city life, and that you just found ways to live with them. This was, then, the ethic that I had adopted about the rats in the backyard—a precarious coexistence that was made easier in my case by the fact that, although I knew there were many rats making their home in the crawl space under the house, I had never seen one *in* the house. So, I thought about the house as *our* space, the crawl space as *their* space, and the yard as shared space.

This balance was disrupted, however, when new neighbors moved in next to us, on the other side from the long-term residents who would later build a new home on their two lots. These new neighbors were a white, middle-class couple who were expecting a baby early the following year, and they had anxiety about the character of the neighborhood. The woman talked frequently about how they had taken a risk moving into this "dangerous" neighborhood (which I think must have been referring to the neighborhood's racial makeup because crime rates were not especially high). But they had had to move to the neighborhood, she explained, because housing prices in Seattle had risen so much that this was the only place they could afford. She frequently called the police on what she determined to be suspicious activity, which encompassed a range of things, from a Black teenager in a hoodie sitting on a curb down the street, to a Hmong elder collecting aluminum cans from a neighbor's recycling bin put out on the curb for garbage pickup, to a Latina woman who lived in a camper parked down the street.

Beyond the perceived danger of the neighborhood, she was also concerned about the fact that people left mattresses, furniture, and other items out on their parking strips and corners to give away—items that sometimes sat there for days or weeks that she thought were filthy and unsightly and which she reported to the city. She complained about the Mexican family across the street because the father and son worked together on fixing up and selling old cars—an activity that meant there were often multiple cars parked on the street in front of our houses. "They're using more than their share of parking," she said to me on more than one occasion, and reported this family's vehicles to the city for not having the proper parking permits—permits that cost sixty-five dollars per vehicle per year, but which were not enforced on our street unless there was a complaint. Every time I talked to her, she had a new grievance about some aspect of life in the neighborhood.

And then one day she sent me an email to let me know that we had become the subject of her latest concern. I had seen and spoken with her earlier that day, and she hadn't mentioned anything, but in the email she cursed at me about the rats she had seen in our yard on multiple occasions, animals she found to be enraging, disgusting, diseased, and dangerous. If we didn't attend to the problem immediately by calling an exterminator, she would report us to the city for not taking steps to eliminate the rodent population. She said that she would give us a week to show that we were addressing the situation, and if we had not, she would file a formal complaint. She believed this was our problem because we had chickens, and she had seen the rats in our yard. *Get rid of the rats*, she warned, *or else*.

Our neighbor's threat to report us (and the rats) to authorities was not made in a vacuum—human intolerance for rats (and other so-called pests) has direct effects on rats' ability to live their lives in shared multispecies spaces, and this ability is curtailed more frequently in gentrifying neighborhoods. The geographers Phil Hubbard and Andrew Brooks explain that "wealthier, whiter populations are . . . more prone to report rat sightings to the authorities than lower status socio-economic groups . . . meaning that efforts to displace rats are more pronounced in gentrifying neighbourhoods than in low-income ones."[12] This was articulated clearly in the marked differences in the responses of the neighbors to the presence of rats: on the one side, among many of the longer-term residents I knew, there was a level of tolerance and acceptance of rats as an inevitable part of multispecies urban life, even if they may not like it; on the other, there was what Hubbard and Brooks refer to as the "'NIMBY' [not in my backyard] anxiety that incoming gentrifiers voice about rats [that] mirrors the anxieties they can express about the other feral or stray urban inhabitants thought to pose a threat to the cleanliness of 'their' neighbourhood."[13] I saw this new neighbor's anxieties about "diseased" and "disgusting" animals like rats, the "filthy" and "unsightly" furniture left on parking strips, the "danger" of the neighborhood as a whole, and the neighbors using "more than their share" of the parking as reflections of her idea that the neighborhood should be a particular type of space—*her kind of space*—and she felt justified policing the neighborhood to try to enforce the kinds of norms she desired, what Hubbard and Brooks describe as a "middle-class version of defensive homeownership."[14]

This also reveals a deeper ignorance about rats—an ignorance many of us share. The idea that rats carry and transmit virulent diseases that pose a threat to humans and species designated as "belonging" in urban spaces (like dogs and cats) is one of the most common negative narratives about these despised urban species. However, most people I've encountered who articulate fear about rats carrying diseases that threaten humans and animals kept as pets don't speak about the issue with any specificity—neither about which diseases rats carry nor about the actual prevalence and level of risk these diseases pose. I, too, felt this anxiety before I looked into the actual threat rats pose to human and animal health.

Like any animal—human or nonhuman—rats can carry diseases, and in the case of rats, these can include leptospirosis, plague, and tularemia. I was curious to better understand the risk posed by rats and these diseases in Seattle—a city with a constantly booming rat population—and so I did some research on the prevalence of the diseases in Washington state. The bacterium *Yersinia pestis*, which causes plague, is not found in wild animals in Washington, so this was not a concern in terms of disease transmission on Beacon Hill.[15] Tularemia can be transmitted through contact with an infected dead animal or bites from an infected animal, and through contaminated water and soil.[16] There are between one and ten cases of tularemia reported annually in Washington.[17] Leptospirosis is a common concern related to rats in Seattle, especially in the context of veterinarians recommending the leptospirosis vaccine for dogs and cats. Transmitted through the contamination of water from infected urine, leptospirosis can cause significant liver and kidney damage. There are between zero and five cases of leptospirosis in humans reported in Washington state every year, most commonly contracted through swimming and other water activities, not through the presence of rats in homes or yards.[18] The risk of contracting leptospirosis is higher in dogs than in humans: In 2016, for instance, there were twenty-two canine cases in Washington; in King County, between 2007 and 2016, there was an average of sixteen canine cases per year—numbers that could be reduced with wider vaccine use.[19]

As I was writing this chapter, an article published in the *New York Times* stated the following: "New York City Rats: They're in the Park, on Your Block and Even at Your Table—Reported rat sightings, health inspections finding evidence of rat activity and cases of a disease spread via rat urine are all up amid the pandemic."[20] The article reports that there had been fifteen cases of leptospirosis so far in 2021 (as of November), including one death, presumed to be contracted through contact with rat urine (although the article does not identify the actual source of the disease in any of the cases). In a city of 8.8 million people long known for its high rat population, fifteen cases of leptospirosis and one death struck me as a surprisingly low number, when compared with the frequency of narratives about rats as a major threat to human health, and indeed the belief that this topic warranted substantive coverage in a major national (and internationally read) news outlet. My intention is not at all to make light of the fact that one person died and fourteen others were made ill by leptospirosis; my point is more to highlight the overblown fear and anxiety around disease transmission resulting from living in close proximity to rats. Perhaps this anxiety is rooted in the long-standing belief that rats caused the spread of the bubonic plague, known as the Black Death, in fourteenth-century Europe and again in subsequent plague epidemics. Recent research, however, has uncovered evidence that the plague was, in fact, spread not by rats but by humans (in particular, by human-specific fleas and body lice).[21]

It's interesting to think about how misconceptions and assumptions, and the persistent instinct to scapegoat nonhuman animals, can be reinforced across hundreds of years (in this case, across at least eight centuries), and how these kinds of misconceptions then justify the ongoing efforts at a wholesale eradication of entire species. It makes me wonder how many other common assumptions based on misinformation about animals pervade our everyday lives, reinforcing the supremacy of humans and excusing routine forms of violence against other animals. What assumptions are we making right now that are causing harm to those whose lives intersect with our own? And how do we so often know so little about someone with whom we live so closely?

After learning about this relatively low risk of disease transmission related to rats, when I noticed the rats out during the day, I also researched what measures could be taken to reduce this small risk of disease transmission and implemented these in the backyard. First, I stopped tossing feed out into the yard for the chickens to find, and instead left their feed in their coop, which was open during the day for them to access but which the rats were less likely to enter. The dogs and cats were vaccinated for leptospirosis (as well as the other recommended vaccinations, like those to protect against bordetella and rabies). And because water is such a common site of disease transmission, and there were several containers as water sources for urban wildlife in the backyard, I made sure to change and scrub these containers daily. It wasn't, of course, creating a sterile, disease-free environment in the backyard—that would be neither possible nor desirable—but I worked to reduce the small risk of disease transmission that the presence of the rats posed.

I explained all of this to the neighbor, but her beliefs about the danger posed by rats seemed to prevent her from being able to consider this information. She again threatened to report us to the authorities. I tend to be a (sometimes unfortunate) rule follower, and I was afraid of getting into trouble. I also didn't want to have ongoing conflict with any of our neighbors. Although I was angry at her hostility about the situation, I also wanted to understand where she was coming from—and I think it was ultimately fear that drove her response. Seeing her behavior as motivated by fear made me feel for her situation.

I bought a live animal trap with the idea of trapping the rats one by one and moving them somewhere else (to the Washington Park Arboretum, I was thinking—which it turns out may be illegal). I hoped to catch one every night and move them in a steady stream to the arboretum, which would provide a beautiful lush new habitat on Lake Washington—nicer, I hoped, than the crawl space. I worried over the separation of kinship networks that would be caused by moving them one at a time to the park, but I thought it was better than killing them, and I hoped that if I dropped them off in the same place, they would somehow find each other in their new home. This was the plan; however, I baited the trap night after night with peanut butter, and each morning the peanut butter was gone—licked clean—and there was no rat in the trap.

At a loss, I started researching exterminators. I was desperate to find a "humane" way to manage the rat population, and so I asked each person I called about their methods of eradicating rats. As I learned from these conversations, there is a large arsenal of methods people use to kill rats, and, as I intuited in this process and as the Humane Society of the United States states plainly, "There are truly no humane ways to kill rodents."[22] Various types of poison are widely used, such as anticoagulants that prevent the blood from being able to clot and kill the animal slowly from the inside. Poisons also pose a risk to other animals and children who might eat the poison and to animals who might eat the poisoned rat. Glue traps are boards coated with a strong glue that the animal gets stuck on; although sometimes the glue can suffocate the animal as they struggle to get free, it is more common for the animals to die of exhaustion, dehydration, or starvation, as they remain stuck to the trap for days. There are also "zap" traps, which electrocute the rat in a box, and which claim a 100 percent kill rate. There are mazelike traps, such as the bucket trap that causes the animals to drown. And then there is the classic snap trap that, if working as intended, breaks the animal's neck and kills them immediately; however, these often badly injure the animal and cause prolonged suffering prior to death. I asked every exterminator whether they could livetrap and release the rats elsewhere, and most of them laughed at me.

One guy, though, surprised me by saying, "Oh, yeah, I can livetrap 'em. No problem at all."

Over the moon, after having resigned myself to the idea that the rats would have to be killed, I replied, "Oh my god, really? And then you release them into the forest or somewhere nice?"

He chuckled, "Naw, I sell them to a research lab." I balked at the thought of these rats, living their lives freely, being sold into captivity to be test subjects. This was long before I knew much about the use of animals in biomedical research, but my gut reaction was one of dread. I told him I'd think about it and call him back.

In retrospect, I find it curious now that there was a research lab interested in buying wild-caught rats; the labs with which I'm familiar today rely on carefully controlled breeding programs and genetic characteristics in their rodent populations to ensure as much consistency and reliability as possible in their protocols. But I wasn't thinking about this then; instead, I was wondering what they would be used for in these research labs and what their lives would be like. I thought about it at length and decided I needed some more information about the kinds of labs this exterminator was talking about, and so I called him back to discuss it further. He didn't answer, and I was never able to get a hold of him again.

Across all the other exterminator calls, the recommended method was to use poison bait because, they explained, it's nearly impossible to reduce such a large number of rats with the other methods of killing. And so, I settled on an exterminator who assured me that the variety of poison he used was the most "humane" form, describing how the rats would eat the bait, start to feel flu-like symptoms, and then just get sleepy before they die. He also said that the poison makes them thirsty, so they seek out water sources, like sewer tunnels, and thus there rarely are any remains of rats for human residents to manage. I suspected that, knowing I was concerned about the well-being of the rats, he was fabricating this notion of "humane" poison and was just trying to sell his services in describing relatively mild-sounding symptoms that rats experienced leading up to their death. Or perhaps this is something he truly believed—something he had to believe in order to do his job? Whatever the reason, at that point I was desperate to believe that there was, in fact, a more humane—or at least less horrible—way to kill the rats, so I chose to believe him. This was a choice. As much as I can try to justify this as an unavoidable action, as necessary, as an unfortunate reality of living in proximity to rats and to human neighbors, it was still a choice to kill these rats. And it was one that went against my better judgment and the nagging awareness that there is no humane way to kill a healthy animal who doesn't want to die.

In thinking about this choice over the years, I've noticed the powerful inclination to believe in fictions about the violence against animals, readily taken up even by those who are normally careful and critical thinkers. These fictions are all around us. Ideas about "humane slaughter" of farmed animals make killing them for the fulfillment of our own desires more palatable. Narratives around "necessity" make killing animals seem like an unfortunate but unavoidable reality. But these are oriented in an anthropocentric worldview—one that is anchored to the belief that ultimately animals are here for us to use, on one side, and to eliminate when they become a nuisance, on the other. Nowhere in these narratives that justify killing is there a foundational consideration of the actual animal, their life, and their kinship relations—or if there is, it is soon eclipsed by the persistence of human exceptionalism.

On the day the exterminator arrived to clean out the crawl space and set up the "bait stations," I felt overwhelmed with dread and rage. Dread at the impending mass killing for which I was responsible through my decision to exterminate the rats. Rage at the neighbor for insisting I engage in this mass killing. I huddled in my office, working on the computer, trying to ignore and forget what was happening beneath the house, and then there was a knock on the door.

"Ma'am," the exterminator said, "you've got a serious infestation—thousands of them. Usually, we see a few nests in the insulation in a crawl space, but your crawl space is one continuous nest. Quite frankly, it's disgusting."

"Oh gosh," I said, feeling like this was a reflection on me—that I was also disgusting, living in a house with so many rats. I followed him outside, peeking into the door to the crawl space, surprised that the exterminators had pulled out all of the insulation from the ceiling. I remember noting how cozy that insulation looked—a perfect nest for a rat.

As I was standing there talking to the man, I suddenly heard a high-pitched cry from the neighbor's yard. It was a piercing sound—an animal in pain. The neighbor had lined the ground along our shared fence with the classic snap rat traps. I followed the sound and saw that there was a rat caught in one of the traps. They must have fled the crawl space when the men were cleaning it out—expelled from their home—and run right into the trap. The rat's back end was caught in the trap, crushed and distorted, their back legs twitching, their front legs scrambling at the air trying to escape. I saw that there was clearly no hope for this rat to recover; they were too badly injured. I ran around to the neighbor's door. I told her there was a live rat stuck in her trap and asked her to do something about it.

"What do you want me to do?" she asked, uneasily.

"I don't know, but you need to do something." I felt sick with my own guilt, angry at the neighbor that I found myself in this situation in the first place, and desperate for the rat's suffering to end. Although the exterminator I had hired was busy killing thousands of rats at my instruction, this individual rat made palpable and visible the stakes of what I had chosen to do.

The exterminator heard what was going on and came around to the neighbor's door, asking if he could help. The neighbor nodded emphatically.

He was holding an aluminum rake he had grabbed from the side yard, went around to the trap, and hit the rat with the rake. He must have struck the rat a dozen times before they died, the animal letting out even louder cries with every strike. Eventually their mangled body was left bloody and lifeless in the trap. He picked up the rat by the tail, trap and all, and tossed them in the garbage can in the driveway. Somehow the suffering and killing of this rat were made even more upsetting by their disposal in the garbage, by the way their body was simply thrown away—a sign of their disposability, their abstraction not as an individual but as part of a despised species, whose erasure and mass killing were normalized and encouraged. How many rats had been killed in Seattle alone this year? In the last hundred years? In the whole country over the last hundred years?

Over the next few weeks, the number of rats declined noticeably. I still saw a rat now and then out during the daytime, but it was a fraction of the number I usually saw. I never saw a dead rat. Not a single one of the thousands the exterminator promised to kill with the poison stations. I could see why this was a popular way to kill rats—there were no screams from a trap in the yard, no watching the rats suffer the slow deaths they die bleeding from the inside, and no remains of dead rats to dispose of. There was simply a swift decline in rat sightings and a fuzzy knowledge, pushed easily to the back of the mind, that they hadn't just "gone away" but had been killed en masse. It was easy to not think about them at all. It was easy to forget what I had done and to go on with daily life, unchanged.

But if we attend to what we so easily forget, if we acknowledge what we have actually done, then we can see that the garden is filled with ghosts. It's the ghost of the rat with the crooked tail whom I had watched with such affection and whom I never saw again. It's the ghosts of the thousands of rats who were that rat's kin. It's the ghosts of all those—human and nonhuman—who were dispossessed, displaced, and killed so that this land could be cleared and housing constructed in the first place. To *forget* can be both an inadvertent neglect and an intentional disregard.[23] To forget something means that you have once known the thing that you're inadvertently or intentionally trying to erase. Forgetting is also a privilege and a luxury; it reflects the possibility for distance from the thing that needs to be forgotten in order to sustain life as usual.

To forget is a disconnection—it requires a severing from the thing that doesn't want to be forgotten. And this leads to alienation from the things we've done, the places we inhabit and shape as our own, the forms of kinship and intimacy that are possible. To remember these things that have been forgotten is to start to do the reparative work of recognizing and realizing our culpability—of realizing a different future. If to *dismember* is to tear apart, then to *remember* is to put back together again, to recall the things that were forgotten, splintered into fragments of pain and violence and death, and to fuse them back together, scars and all, in a process of repair, of healing.

I didn't want to write this chapter. I didn't want to think about my culpability in so much violence. I've worked hard over the past decade or more to try to forget this experience, push it to the back of my mind, to not think about it at all. When I do think about it, my instinct is to try to shift the responsibility for all of this killing onto the neighbor who demanded I exterminate the rats. Still, years later, I feel a strong pull to blame her, to try in some way to absolve myself of responsibility for the elimination of the rats with whom we had, until that point, coexisted. It's uncomfortable—no, excruciating—to face the shame of precipitating such extreme violence, to acknowledge the insidious forms of anthropocentrism that make it possible to prioritize the comfort and norms rooted in human desire over the very existence of a nonhuman someone else (and thousands of nonhuman someone elses).

   I also didn't want to write this chapter because it meant reckoning in a public way with my implicatedness as a white person, settler descendent, and gentrifier in the ongoing harm and devastation of white supremacy, settler colonialism, and gentrification for other humans and nonhuman animals. Since moving to Beacon Hill, I have thought at length, privately, about my lack of awareness and thought (perhaps more likely willful ignorance?) in the decision to buy a house in the neighborhood. At the time, I didn't think about how my moving into the neighborhood would change the place itself or that it could inhibit the ability of those already living there (both humans and nonhumans) to stay in their homes. My sense of entitlement was strong, and I felt lucky to have found such a wonderful place to live that was within an achievable price range, even as housing prices in other parts of Seattle had already climbed well above the realm of affordability. I felt a sense of having "discovered" a secret gem in Beacon Hill, and I encouraged friends who were also looking for a more affordable home to check out the area. I didn't think about my *not belonging* in this space because I made the taken-for-granted assumption that I belong here. Wherever we go, we impose, at best, this belief that we *do belong* and, at worst, that those who were already there *don't belong*. This is true in terms of notions of belonging/unbelonging of humans, and it is also true of other species.

Belonging is an exclusionary project. Belonging is often defined through articulations of who *does not* belong, and then those designated as not belonging become marked for expulsion and erasure. Hubbard and Brooks wonder whether or not rats and other similarly maligned animals, such as cockroaches, flies, and bedbugs, "could ever be considered good neighbours given their capacity to crawl and scuttle around domestic interiors biting, eating and defecating. Some animals, it appears, lack the capacity to be seen as anything other than disturbing and 'out of place' in the city."[24] Can rats ever be seen as belonging? Can they ever be treated as coresidents, as urban kin, as part of a vibrant multispecies urban community?

In projects of "greening" cities, there is a recent move to acknowledge that cities are multispecies spaces and that urban planning ought to include designs to help certain species flourish. These efforts manifest in different ways, such as intentionally (or unintentionally) creating nesting sites for birds of prey; restoring urban wetlands and the health of local waterways; and creating space for beekeeping and planting the landscape with native, pollinator-friendly plants, among other efforts. Especially during the COVID-19 pandemic, when many people were working from home, there was a sharp increase in bird-watching, feeding birds, and appreciation for certain kinds of urban wildlife—the welcome kinds of birds: finches, Steller's jays, woodpeckers, chickadees, and others. At the same time, urban infrastructure has become more hostile to other types of animals—evidenced perhaps no more visibly than in the case of "bird spikes"—sharp metal spikes installed on roofs, signs, and other urban infrastructure with the express purpose of preventing pigeons, crows, seagulls, and other undesirable birds from roosting, nesting, and defecating on whatever is below. There is a deepening chasm between belonging and unbelonging among urban animal species—efforts to include as much as to exclude.

Is it possible to transform these deeply ingrained notions of entitlement that create and destroy the conditions for other animals' flourishing—even, and perhaps especially, for those who have been the consistent target of erasure? There is a recent effort by some scholars to think about the productive potential of making animals, especially wild animals, property owners—extending to these species the property rights and protection that come with human ownership of land. This proposal occurs in the context of trying to address the dispossession and harm that come from animals not having the right to live and be safe from expulsion from the land they inhabit. One of the meanings of *belonging* is in the sense of property—that land as property or an animal as property *belongs* to someone, and thus can be subjected to all of the indignities and violence that come with being property. The idea here is that by being made into property *owners*, animals can be insulated from the worst effects of being property themselves. But this move does not undo the foundational violence that comes with colonial framings of property relations. It is the very notion of property—of something or someone else *belonging* to someone—that is a key part of the problem.

If we think about *belonging* in terms of ownership, how do we even know who possesses ownership when there is a vast landscape of different species and lives that need collective access to land they call their own? Might we think about public land, for instance, as shared space—a commons? How do we think about privately owned land, like the backyard, as belonging to a multispecies web of life—as land not belonging to us through an exchange of money and a piece of paper that says we own it, but perhaps as land that we share, that we inhabit with others for the little while that we are alive and that we live on as gently as possible? As pattrice jones explains, "At the heart of the problem is alienation, separation, dissociation. To imagine that you 'own' a piece of land, you must first alienate yourself from it, psychologically tearing yourself out of the seamless fabric of your ecosystem in order to lay claim to part of it."[25] Perhaps, then, expanding ownership rights is not the answer.

In trying to understand the way gentrification functions in relation to belonging, Matthew L. Schuerman argues, in his book *Newcomers: Gentrification and Its Discontents*, "It is hard to argue that a neighborhood *belongs* to one people or another. Many of today's gentrified neighborhoods were once built for the gentry of the nineteenth century, fell from fashion in the mid-twentieth century, and have become desirable again in the past few decades. Other gentrified neighborhoods descend from working-class roots but have gone through considerable ethnic changes."[26] This raises a perennial question about who was here first. If we look only at recent histories of human occupancy, for instance, we might see that Beacon Hill was, for decades, a neighborhood with an especially high percentage of Asian and Pacific Islander American residents, and an overall majority of people of color, and that the incursion of white gentrifiers into the neighborhood has displaced these earlier residents.[27] If we track back further into history, it was Coast Salish peoples and diverse nonhuman species who were here first and consequently dispossessed and displaced (although is it important in this historicization to not erase those who are still here today). If we look back in history even further to a prehuman era and the mass extinction of nonhuman life that followed, the question of who was here first becomes even more complex. So perhaps this question of "who was here first" is only one of the questions we need to ask. Perhaps we need a different understanding of belonging. Perhaps the problem is not entirely about the fact that we, who are settler descendants, are *present*, so much as it is about *how* we are present—that the very foundations of our presence have been and continue to be oriented in and through colonial frameworks of property, occupation, the hierarchical ordering of life, an anthropocentric worldview, a relation to the world as exploitable resource and as capital and ourselves as entitled to the spaces we inhabit. How could this be different?

As I discussed earlier in the book, decolonization is in large part a land-based project—one that repatriates land to those Indigenous communities to which the land originally belonged (not in a colonial property sense of the word but in a relational form of multispecies care, kinship, and stewardship). We might think about the process of recrafting notions of belonging in the city as a decolonial project. With this as a starting point, how do we think about land occupied now, belonging in an ownership sense to settler descendants? If those of us who are settler descendants returned the land we think we own to Indigenous communities and to all of the nonhuman species who have been harmed by this ownership, what kind of futures are possible?

The LANDBACK active movement transfers land back to Indigenous communities, as has been successfully accomplished in cases like the transfer of Duluwat Island in Humboldt Bay to the Wiyot people by the city of Eureka, California; or the transfer of management rights of the Bison Range on the Flathead Indian Reservation to the Confederated Salish and Kootenai Tribes in Montana by the US Fish and Wildlife Service; or even the purchase of the Salish Lodge and surrounding area in Washington state, including Snoqualmie Falls, by the Snoqualmie Tribe (from the Muckleshoot Tribe) to restore important ancestral land to the Snoqualmie people. Yet another example, local to the Puget Sound area, that works to recognize Indigenous sovereignty is the Real Rent program, through which residents of Seattle and the surrounding area can pay rent to the Duwamish to support community and cultural activities and tribal services. This program asks less of settler residents of the area in the sense that it doesn't ask for land to be given up but instead asks for an acknowledgment through monetary support that the land is theirs and that it can be rented.

When I've engaged in conversations with other settler descendants about the repatriation of occupied lands to Indigenous peoples, one of the questions that immediately comes up is, *but where would we go?* I've felt this anxiety myself in trying to think seriously about decolonization and what it would require in real, action-oriented terms. There's a way that the asking of this question and the fear and anxiety it provokes in thinking seriously about an answer cause a persistent inaction—first, a "feeling better" that we are thinking and talking about it at all, soon followed by an uneasy acceptance that there is no alternative, that we're here and there is nowhere else to go. But this discussion and thought process (and, most important, the belief that there is nothing to do about it) reflect the colonial way of thinking that pervades contemporary and historical land-based politics—the belief that if we don't own land, then we don't have a right to inhabit it, that we would have to leave. This way of relating to the land is so deeply rooted in an inability to see land as anything other than property, and belonging as anything other than a project of exclusion, that we can't do anything but assume that if the relationship of ownership and belonging were reversed—if Indigenous communities and nonhuman animals once again had full sovereignty over their ancestral lands—settler descendants would, naturally, be expelled. Perhaps we would be. Or perhaps we wouldn't—perhaps an entirely different way of living and belonging is possible if we let go of the ways of relating to each other, to other animals, and to the land that have, for so long, done so much irreconcilable harm.

Repatriation of land to Indigenous communities by settler subjects is not the only way through which anticolonial futures might be made possible. The forms of knowledge that are created and the systems of education out of which this knowledge emerges are core to transforming colonial (and capitalist) relations of harm. Leanne Betasamosake Simpson, in describing the importance of Nishnaabeg knowledge, writes that we must "rebel against the permanence of settler colonial reality and not just 'dream alternative realities' but to create them, on the ground in the physical world in spite of being occupied. If we accept colonial permanence, then our rebellion can only take place within settler colonial thought and reality."[28] Simpson articulates methods of education oriented through Nishnaabeg ways of knowing through the land. She prompts questions about what it means to educate ourselves outside of and against colonial systems of knowledge. These are place- and culture-specific practices of learning.

In her book *Undrowned: Black Feminist Lessons from Marine Mammals*, Alexis Pauline Gumbs talks about a fundamental reshaping of how we learn, how we come to know the world, and fundamentally how we might think about the concept, practice, and aims of what "school" is and could be:

> What if school, as we used it on a daily basis, signaled not the name of a process or institution through which we could be indoctrinated, not a structure through which social capital was grasped and policed, but something more organic, like a scale of care. What if school was the scale at which we could care for each other and move together. In my view, at this moment in history, that is really what we need to learn most urgently. . . . I am wondering if we could trade the image of "family" for the practice of school, a unit of care where we are learning and re-learning how to honor each other, how to go deep, how to take turns, how to find nourishing light again and again . . . what if we just need to go back school?[29]

What might it look like to cultivate ways of learning and knowing outside of settler structures within which we are so embedded? What would it look like to "go back to school" to reshape how we think about the cultivation of knowledge and our place in these webs of multispecies relations? How can careful attentiveness to other species, ecologies, and the land itself as they exist on their own terms help to manifest these other forms of knowledge? This is the work I've aimed to do in each of the encounters I describe in this book. In the questions I've asked and the stories I've told, I hope to have offered ways of knowing differently and fostering sensitivity to less harmful forms of relating to others as we work to build different futures.

As I'm thinking about these futures, I'm imagining an annihilation of the settler subject and the human supremacist. This doesn't mean that those of us who are settler descendants or humans must cease to exist or that we must forget or try to obscure the harm that we have done. Rather, it's a dismantling of the settler subject and human supremacist as agents of future harm. In an anti-anthropocentric and anticolonial future, there is no settler subject, no human supremacist, if these logics of settler colonialism and anthropocentrism can be undone. Who are we and who can we be outside of these forms of subjectivity? How can being human mean something different, something better—a radically gentler way of inhabiting our bodies and our lives? And what might that look like?

In her book *Against Purity*, Alexis Shotwell observes, "It is hard for us to examine our connection with unbearable pasts with which we might reckon better, our implication in impossibly complex presents through which we might craft different modes of response, and our aspirations for different futures toward which we might shape different worlds-yet-to-come."[30] I want to return again to this notion of belonging that has haunted the stories I've told—the costs of not belonging for species as despised as rats and for others, and the possibilities for different understandings and practices of belonging. Belonging is, in its final iteration here, about crafting notions of home. It's about a sense of attachment, a sense of relationality, a sense of place. As social beings, human and nonhuman, we crave a sense of belonging to a greater whole—to a place, to a people, to a web of multispecies relations. Can we let go of the ways we alienate and are alienated from each other, and the practices of harm that come with feelings of human entitlement, in order to shape a future where *belonging* is enacted in its fullest sense, its sense of a creation of kinship, of home?

6

On Cows in the Woods

I live in an enchanted forest. So do you. The forest is on
fire. Eros can save us, but only if we are willing to forswear
pride in order to rejoin the joyful worldwide resistance
against humdrum human hegemony.

—pattrice jones, "Queer Eros in the Enchanted Forest"

On my first morning visiting the rural Washington town where my dad and
uncle lived, I woke to shouting from behind the house and the sounds of
thundering hooves. I rushed to the back door and watched as three cowboys and one cowgirl on horseback drove a herd of cows across the back of
the property. They were struggling to get the cows to go through a narrow
gateway that would allow them access to the main road, across which they
would eventually be driven to reach grazing land on the other side of the
valley. The herd was made up of cows with milk-filled udders, their calves,
and young steers. The herd was in chaos, panicking and running in multiple directions, clouds of dust billowing up from the parched earth. One
of the men shouted, "Get in there, you sons of bitches! You know where
to go—don't act like you don't! Ya ya! You stupid piece of shit—Go! Ya!
Ya! Ya!" He pulled hard on the reins of the horse he was riding to redirect
her to stop several cows from turning back in the wrong direction. The

horse leapt back onto her hind legs with the jerk of the reins and lunged forward as the rider kicked hard into her sides. The cows responded in kind, leaping away from the horse and rider, their bodies crashing into others. A calf tripped and pitched forward, narrowly missing being trampled as he righted himself just in time.

Despite the frenzied speed of movement in the herd, progress moving the cows through the gate was slow as they continued to scatter and try to turn away from the gate. The frustration among the riders was growing. And so, the lone woman on horseback tried a different approach: "Come on, sweet pea, come on, hon, this way! Get goin'! That way! Attaboy! Attagirl! Aren't you good? Go on now—git! That's right, hon!"

Against the cows' harsh treatment by the cowboy, I felt a flicker of relief at the gentle nature with which she talked to the cows, but then I was jarred by the differences both in the way they talked to the cows and in the way I responded to it emotionally. It felt similar to how I felt when we had decided to raise chickens for eggs—relief at the feeling that these animals were somehow being treated better, more humanely. I thought about how notions of care and kindness create particular narratives that can soften or obscure the harm of our everyday encounters with animals, like the cows in front of me who were being raised for beef.

With the cowgirl's gentler coaxing, paired with the rougher force of the cowboys, the cows were eventually herded successfully through the gate. Past the gate, they were driven down the dirt road, across the one-lane highway in front of the house, and to the other side of the valley, leaving behind only a flurry of hoofprints and trampled manure on the dusty ground.

What are the experiences of these cows on the move? How might we think critically about the restrictions on their movement, their lives, their social relations, and eventually their deaths that are all so integral to farming animals for meat, dairy, and eggs? What does a freer ranging life allow for, and what possibilities for living and flourishing does it foreclose? How is the destiny of farmed animals predetermined from birth to death, and what would it look like, instead, for these animals to be free from being farmed, allowed to live self-determined lives of flourishing?

The tiny town where I had woken that morning, like many small towns in the region, is situated in the afterlife of the boom-and-bust gold mining of the late 1800s and early 1900s. Mining and homesteading were made possible through war and the seizure of unceded land from Indigenous tribes who were the original inhabitants of the region. In the 1870s, local tribes were violently dispossessed of their land and confined to reservations authorized by the federal government. Land seizure, dispossession, and the development of reservations meant that miners and settlers were free to build mining outposts to extract what gold they could from the area. In the late 1800s, mining proliferated and the construction of frontier towns occurred rapidly, including hotels, saloons, general stores, brothels, banks, post offices, assay offices, and other businesses oriented around supporting a frenzied trade in gold. Quickly, though, gold resources dwindled or became too expensive to extract, and the towns throughout the region began to decline. Although mining continued in the area into the 1900s, peak populations shrank as many people left in search of other opportunities. Some miners and settlers stayed and took up agricultural pursuits, but the mining towns became shadows of what they once had been.

Today, there are the remnants of frontier log cabins decaying into the landscape; others are gone entirely, deconstructed and scavenged by people in the area repurposing wood for their own building projects. Some remaining clusters of original structures are marketed to tourists as ghost towns, calling up the familiar legend of boom-and-bust gold mining towns abandoned and left to decay. The romanticism of ghost town lore, though, can have the effect of obscuring the contemporary dynamics of the vibrant community of very-much-alive people and animals and the surrounding hills and valleys. Life is unfolding in this place, much as in other places, with the humdrum of the day to day.

The humans in the area are an eclectic bunch—conservatives, liberals, people who don't align easily with any political perspective or are solidly antigovernment—ranchers, aging back-to-the-lander hippies, people whose families settled there long ago during the town's gold mining days or who have come in more recent generations to live an off-grid simpler life away from the bustle of cities and suburbs or who enjoy the distance from more overt oversight of the law. In a strange way, it's a sort of lawless place with the flavor of the Old West.

Although the community encompasses many different perspectives and lifeways, visiting from the liberal enclave of Seattle, what was most noticeable to me was the vein of far-right conservativism in the area. Locals in pickup trucks pass through, stopping for beers at one of the taverns or continuing on, some sporting Trump flags and bumper stickers that read "Take America Back." Because of the region's proximity to the Canadian border, border patrol polices the area—a reminder, I thought, that *America*, as a project of continual nation-making, has to be maintained as much in its policing of borders (what *is* or *is not* geographically *America*) as in the demonstration of patriotism by those who consider themselves to be proudly *American*.

Beyond and entangled with the human community, there are cows, horses, and sheep who graze in the fields. There are cats lounging on porches, rabbits raised for meat huddled quietly in cages under a tarp at the neighbor's house across the road. There are families of deer grazing together on the side of the road, darting into underbrush when a car comes too close. There are bluebirds, red-winged blackbirds, orioles, magpies, finches, and robins darting through the air, landing on wire fencing near the wooden bluebird boxes my uncle helped to build and install in the area. There are hawks, eagles, and owls who hunt, soaring high and then swooping down, talons extended to grab a mouse or chipmunk from the ground. I wondered what interactions these animals have with each other, and whether they feel a sense of community sharing this place.

I'm certain, of course, that I am missing almost all of the nuance of the communities that make up the region—both human and nonhuman. I left Seattle and moved to this town a couple of years after my first visit when I witnessed the roundup of the cows, and although I've now been living here three years, my newness to the area makes it difficult to really know the place or to understand its lived social dynamics that unfold over decades (and millennia). It's also hard to ever understand what it is to inhabit another human's or animal's lifeworld. But an effort to try, to fail at understanding, to try again, to fail, and to try again gets us closer to, if not understanding, then at least a sensitivity and attentiveness to difference and the ways we might honor that.

It's an attentiveness, too, to ghosts, to haunting, to acts of violence and the possibility of repair. Attending to the history of this place cultivates a sensitivity to the fact that, even in the vibrancy of life, there's an eeriness to the landscape—echoes of violence, of struggle, of dispossession and abandonment. There's a sense of danger in its remoteness and its lawlessness, a darkness that can't always be pinpointed to a source. But there's also beauty and meaning made in intimate relationships of care and fleeting moments of mysterious encounter.

As I've tried to understand this complexity, I've wondered: What if we hold in balance that this place today is simultaneously a vibrant community and a place full of ghosts? What ghosts walk down the main street or ride on horseback up the trails through the hills? What spectral horses are hitched to posts outside the general store or the abandoned assay office? What long-dead cows graze across the valley? What ghosts of deer and mountain lions drink from the creek? What shadows of owls fly through the night hunting for ghostly prey? Can today's humans, horses, cows, deer, mountain lions, and owls feel their lingering presence? Are they haunted there by the violence that unfolded to build this mining outpost? Might we also think about being haunted by the resonance of multispecies forms of love, care, and kinship that unfolded in everyday life?

The cows I saw thundering past on that first morning were the property of ranchers raising them for beef in cow-calf operations. This was a model of animal agriculture I had heard about many times but had never witnessed in such detail before. My research had thus far been on cows raised for dairy, where production practices rely on the confinement of the animals to varying degrees, and the separation of the cows and calves immediately after birth is core to maximizing profit in an industry dependent on diverting milk as its core commodity into the market. Cow-calf operations run on a different logic, oriented as they are around the production of beef and not milk. As the name suggests, herds are composed of cows and their calves, and these herds roam across ranges, grazing and foraging for food during nonwinter months, either until their young are at slaughter weight (at one or two years old) or until they are transferred to feedlots for longer-term fattening before slaughter. Cow-calf operations involve many of the same practices as dairy and confinement beef farming—branding, castration, breeding, familial separation, transport, and slaughter—but the more benign veneer of cows grazing on the landscape works to obscure the harms involved in these practices.

As I explored earlier in the book in the case of chickens, animal agriculture relies fundamentally on co-opting the reproductive process. Cow-calf operations are no different. What you see when you drive by cows grazing with their calves on the range requires close reproductive planning and management on the part of the rancher.

One of the major differences between dairy and beef operations is that calves remain with their mothers in cow-calf ranching for much longer than in dairy production. When I first saw these herds, I thought about my experience of the dairy industry and the immediate separation of cows from their calves. I thought about how it seemed like a gift that these cows I was seeing now got to spend more time with their young. But dairy science has shown that the longer a cow and her calf are allowed to stay together, the stronger the bond and the more traumatic it is to separate them.[1] In cow-calf operations, the cow and calf have the opportunity to bond for an extended period (often up to two years), both with each other and with other members of the herd. These kinship relationships, then, are severed suddenly when the calf is determined to be ready for fattening or slaughter, fracturing the social connections in the herd at large.

It is usually in the fall, when the weather begins to turn cold, that this separation occurs. The herds are rounded up by humans on horseback, driven into transport trailers, and delivered to one of a few destinations. Females who are still reproductively viable will return to the ranch to enter another cycle of breeding, birthing, and raising a calf. Some calves might stay with their mothers for another year. Some might go to the feedlot if they are young steers in need of fattening before slaughter. Some ranches send calves to feedlots at eight months of age rather than keeping them with the herd for another year. Others might be sent directly to an auction or slaughterhouse if they're determined to be in a state to yield adequate volumes of meat.

Relying so heavily on grazing, cow-calf operations use an enormous amount of land, and much of this land in the region, it turns out, is public. When herds have outgrazed the pastures owned by the ranch, ranchers sometimes lease rangeland from other landowners in the area where they will graze for the season. But many herds are simply set loose on leased public rangelands. The state subsidizes the cost of animal agriculture not only through meat, dairy, and grain subsidies for farmers but also through leasing public land at low prices to offset the costs of feeding and caring for the cows. These rangelands are bounded by cattle guards on the roads: a series of metal pipes or beams spaced across a deep ditch, creating a slotted section of the road that cows cannot cross. But within these boundaries, they have free range of the forest and meadows, and the damage they do to these ecosystems already strained by drought, wildfires, and human development is significant.

During my time with my dad that summer, we took scenic drives nearly every day to explore the region. He had moved from my home in Seattle to his brother's during the first spike in COVID-19 in early 2020, when we thought he would be safer in a less populated place. The living situation suited him and his brother, who had been living alone since my aunt died several years earlier, and so he decided to stay there. My dad didn't drive anymore and had seen very little of the area. We discovered that within just a one- or two-hour drive in every direction there was incredibly varied and awe-inspiring scenery. We saw rivers and lakes, forests and desert landscapes of sagebrush, epic panoramas of mountains and valleys, hilltops and canyons, deer and a bear trundling along across a field. And we saw cows.

I was surprised the first time we encountered cows in the forest. We had just taken a short walk in the woods to see a pair of nine-hundred-year-old towering tamarack trees and had rested on a bench with our necks craned back, looking up into the canopy. To live for nine hundred years and stand there, bearing witness to all that unfolded in that time, was something to ponder. Shortly after we returned to the car and continued on our way, I brought the car to a crawl as we passed a young steer and his mother on the side of the road. We looked beyond them, into the woods, and there was a herd of cows and their calves, and a few deer standing nearby. As we slowed down, they looked at us warily. We drove on. Over the course of our daily drives, we saw dozens and dozens of cows in the woods—some standing in the road, others standing or lying down beneath the trees, drinking from the many lakes in the area, and grazing on patches of greenery. They all had ear tags, denoting the particular ranch they belonged to, and large brands emblazoned on their sides—they were property after all, and property must be marked so as not to be mistaken as not-owned. In this way, ranchers appeared to not be worried about the cows disappearing; in essence, they were bounded not only by the cattle guards that kept them geographically contained but also by these markers of ownership that kept them conceptually and categorically contained— both as farmed animals generally and as the specific property of individual farmers. Everyone in the area knew whose cows these were—those with the yellow ear tags were owned by one major rancher, and those with the red by another. There was no escaping the reality of this configuration of ownership.

I had encountered cows in the woods previously at VINE Sanctuary in Vermont, where the kinds of lives that were possible in those woods were very different. There is wide variation in how sanctuary spaces and forms of care are understood and shaped, but at VINE, one of the core guiding values at the sanctuary is creating the conditions for self-determination by the animal residents in how they cocreate their shared lives, who they will choose as kin, and what contact they wish to have with humans at the sanctuary. There is a herd of cows there who live in what the sanctuary staff call the "back pasture" who prefer to not have contact with humans, instead enjoying the company of bovine kin and other wild animals who pass through and inhabit the sanctuary grounds. It is a part of the sanctuary that humans tend to leave alone, except to check daily that everyone is safe and well. Other cows, who enjoy socializing with humans, choose to live in "the commons" (a multispecies space toward the front of the sanctuary), where contact with humans and other sanctuary residents is frequent and routine. The cows in the back pasture have been allowed to become semiferal and to live out their lives in the ways they choose, albeit within the boundaries of the sanctuary grounds.

The cofounder of VINE, pattrice jones, tells the story of a cow and her calf who made their way from a beef farm to a very different future at VINE:

> One day, a cow jumped over a beef farm fence to birth a calf in the forest, far from the grasping hands of humans. She and her son then found their way to a friendly person who conveyed them to a sanctuary. The mother's fierceness in protecting her son from perceived threats made them a poor match for a sanctuary offering tours to the public, and so they both came to VINE Sanctuary, where they eventually joined the hardy herd in our back pasture. The cows in that community organize their own affairs as they see fit. They often choose to sleep in the forest rather than in the barn and to drink from a brook or pond rather than water troughs. Other than eyeballing everybody twice each day, just to ensure that nobody is ill or injured in any way, sanctuary staff stay out of the way.
>
> Jan and her calf Justin have flourished in that setting. As Justin has grown up into a sweet-tempered young adult with a fondness for bird-watching, Jan has made friends with cows her own age. She no longer glares and prepares to charge any person who might dare to look too lingeringly at her son, but she still becomes visibly wary when strangers appear.
>
> I've endured more than a few uncomfortable moments under the searchlight of Jan's gaze, hoping she will see that she need not charge at me to protect herself or her son. At such moments, it seems to me that she is both mad and mystified, angered and confused by what she has seen people do to cows. In Jan's expression when she looks at

people, even when she is comparatively relaxed, I perceive a combination of challenge and question, as if she is prepared to fight an enemy she cannot fathom.[2]

Although they easily could be like Jan and Justin, these cows in the Washington woods were not those cows. These cows, before and after their time in these woods, are caught up in the web of being bred, raised, and killed for food. The cows in these woods seemed simultaneously out of place and in their place. I thought about how these cows could be absolutely at home there in the woods, living their lives apart from humans in community with other bovines and wild animals. In some ways, nothing felt more natural than these cows wandering the woods.

But at the same time, they seemed jarringly out of place. In *The Oxen at the Intersection*, jones writes about the powerful narratives that define who are the original inhabitants of a place and what those places mean. Once the kind of place it is has been established, then "everything depends on keeping [farmed animals] in their place."[3] And so this is a different "in their place"—not in the sense of their finding themselves in a place to live that allows for the expression of perhaps their fullest selves. But as farmed individuals, these cows in the woods, then, are kept in their place—entrapped in agriculture, their lives and bodies caught up in the cycle of commodity production. And they are out of place, reminders—if you attend to the history—of whose disappearance was necessitated by their presence, of who their presence erased, and of their ancestors dispossessed of their traditional lands and relocated here.

The surprising nature of the cows in the woods created a fractured reality, a signal that something is askew—that cows are not the kinds of animals you imagine seeing in the woods of central Washington. Ideas about where cows *do* belong might creep in, with images of pastures, fields, red barns, and the like. It was precisely these images that made me so delighted to see these cows in the woods, to see them living in a different kind of place, in different kinds of conditions. In the imaginary of meat eaters and supporters of free-range beef, this is the point—*this* farming (small-scale, free-range) is not *that* farming (industrial, high confinement). Free-range beef farming, and especially the practice of regenerative grazing, is being looked to as the especially high environmental impacts of beef production become harder to ignore in the global climate crisis. Clinging to a future in which beef can be consumed with an environmentally clean conscience, an imaginary of free-range agriculture as an alternative is carefully constructed—this form of agriculture not only does not harm the environment, so the narrative goes, but it in fact helps the environment. But the environmental costs of beef production in any form—as well as the production of other animal-based foods—are high, with free-range farming causing damage to fragile ecosystems; polluting land, air, and waterways; and deforesting land for grazing.[4] Seeing the cows in the woods, then, meant considering the environmental impacts of their out-of-placeness in this landscape.

The presence of cows in the woods also renders reality askew in a different way, in the sense of cracking open an imaginary of cows living in multispecies communities of care in the woods, free of being farmed—an imaginary made reality at VINE. As we encountered these cows in the woods, the cows in the woods at VINE were there, in the back of my mind.

It's important, I think, to attend to what it means to be "in their place" and "out of place." This out-of-placeness is all over the region if you are attuned to noticing it. Across the landscape lie thousands of erratics, scattered across fields and hillsides. Erratics are rocks in the landscape that differ in type and size from those native to the area. They are rocks that have been delivered to their resting spot by glaciers in the last Ice Age. They are both out of place and out of time, remnants of a different geological epoch, reminding us or warning us, if we listen, of the changing climate and the perhaps terrifying/perhaps comforting promise of a radically different future—a future that may or may not include humans. These erratics are recognizable *because* they are out of place—a giant boulder in a field of hay, or a massive rock in the middle of the river around which the waters curve and flow.

If we anchor our attention to the out-of-placeness of these erratics, we can see the out-of-placeness of so many other things. There are white settler descendants who now populate the region in what was formerly the home of Indigenous human communities and free-living animal species. The cows are out of place in the woods and in the fields. Horses waiting quietly in their corrals are out of place—horses whose ancestors ran freely in herds throughout the West but who are now tamed, saddled, and bridled, alienated from generations of equine kin, and employed in keeping cows in their place, obscuring the ways they all are out of place. Settler structures are out of place—the barns and fencing and farmhouses. Some abandoned structures are being disappeared by rain and snow and wind and time. These barns, houses, and whole tiny towns from the 1800s and early 1900s were abandoned as the mines extracted their final treasures, their walls leaning and creaking in the wind before they eventually crumple to the ground, a pile of weathered wood.

Some colonial structures are not so easily erased by weather and time. Human-made constructions such as dams create deep wounds in the earth and water, dramatically altering the landscape and destroying existing ecological processes and multispecies relationships unfolding in and around waterways. There are dams all over the region, some bigger than others, some actively producing energy and others sitting, defunct, in rivers that might find new ways to flourish without them. On one of our outings, we went to see one of the dams that was a century old and now defunct. We had trouble finding the turnoff to the bumpy and overgrown dirt road down to the dam, and we ended up instead on a long dead-end side road that passed an abandoned mining shaft. Up ahead we saw a mound in the road, and as we got closer, we saw that it was a rattlesnake. At first, I thought they were sunning themselves on the warm terrain of the road, but upon further inspection, I discovered they were dead and assumed they had been run over by a vehicle. I got out of the car to inspect the snake more carefully, only to find that they hadn't been hit by a car at all, but instead only their head had been crushed, the rattle at the end of their tail cut off. Everywhere there were the signs of human power and violence against the nonhuman world. The dead snake in the road was one of these signs. The dam was another.

My uncle and late aunt worked for decades in solidarity with members of the regional tribes to advocate for the removal of the dam, and I wanted to see the place they had worked so hard for so long to restore. The dam, built in 1920, today generates no power and has no functional use to the state. However, the state has decided the dam's removal would be too costly, and so it sits there, a symbol of human hubris, I thought, as I stood looking down at it.

The removal of dams can, in many cases, have immediate effects in beginning to restore multispecies lifeways. The Elwha River dam removal project on the Olympic Peninsula in western Washington is one example of rapid ecological recovery, where, in 2011, destruction began on the two dams that blocked the natural flow of the river. Led by the Lower Elwha Klallam Tribe, the restoration project has been considered one of the most successful dam removal projects in the world and shows that some environmental destruction can be undone, that the harm caused by the dam can be rendered a haunting memory against which the flourishing of animals and humans unfolds.

A dam is a disjuncture in the landscape—physically, in that it splits the river in two, and ecologically as it disjoints natural ecosystems, processes, and flows of life. Each spawning season, the salmon in the region swim up the river to spawn, their traditional spawning grounds located upstream from the dam. Each season they are unable to reach those waters, and when they are stopped by the large dam, they jump and jump and jump at the fifty-four-foot wall of water, eventually either giving up and spawning downstream or dying of exhaustion at the base of the dam and washing down the river. I think about that knowledge passed down, buried deep in those salmon's genes, that leads them, generation after generation, up the river to spawn. Over a hundred years of being stopped by a dam, there seems to be an ingrained faith—maybe a genetically coded hopeful effort—to finally reach their traditional spawning grounds—knowledge passed through generations that this is *their* place, the place where their eggs will be laid and their young born, perhaps a belief that this place, someday, might once again be theirs.

It was 2022 when Eric and I divorced and I moved to this tiny town to build a new life. Being in this new place, I've thought about the social norms and conditioning involved in how we can and do know animals that are shaped by the places we live, the communities we're a part of, and the work that we do. Inhabiting cities for most of my life, my encounters with other animals were limited to the kinds of experiences I've shared so far in this book, and they reflect a particular situatedness as an urban dweller and consumer. Living now in a rural farming community is different, and my curiosity has shifted to the ways people in rural areas, and specifically those involved in animal agriculture, might think about animals in ways different from my own.

The smaller scale of animal agriculture here means that for many, farmers and ranchers are just trying to get by in an industry that's transformed with industrialization and pushed out countless small farmers. Many of the farms and ranches have been operating for generations, defining in fundamental ways the identities of these farming families and the broader social dynamics of the area. Financial constraints shape the lives of farmers and ranchers, and they affect the lives of the animals being farmed in terms of the care they can receive. In this socioeconomic context, some of the continued commitment to farming is motivated by the need to make a living in a place with few other employment opportunities, making agriculture a much-needed source of income, even as it becomes harder to get by in this line of work.

Living in such a remote rural area where many people are living below the poverty line constrains in some ways how people can care for animals (farmed or not). This extends to forms of care like how animals are raised, how they are fed, and what medical care (if any) they can receive. My neighbor, for instance, keeps a flock of sheep for whom she doesn't (or isn't able to) provide food on a regular basis. In the summer, they eat the grass in their pasture down to the dirt, and each winter, multiple sheep die of starvation in the deep snow. When they're sick, they either live or die on their own. Veterinary care can be geographically and financially prohibitive, or it may not even be a consideration for animals whose cost of care is calculated against the income they can bring in as commodities.

Contact with farmed animals in communities like this one is much more direct than the more removed relationships urbanites have with animals who arrive on their plates as food. Whereas social conditioning for people not involved in farming operates through the normalization of an abstracted notion of animals as food and the consumption practices that follow from it, the conditioning of farmers and ranchers involves another layer of socialization in what it means to raise animals for meat, dairy, and eggs. Although rural and urban communities differ in their connection to these animals, the same fundamental beliefs that animals are food are present in both. Familiar with the enculturation of urban dwellers, I was interested in how people involved in animal agriculture are socialized into farming. Some of my experiences here in this place so far have offered a window into this world.

In the spring after I arrived, when all the calves bred for beef had been born, the ranchers and hired workers at the largest ranch in the area performed the castration and branding of the calves. They make this event into an anticipated party—with people socializing, laughing, bringing beer and food, and coming together to make this shared seasonal task enjoyable. One of the men I know (I'll call him Dave) is hired as a temporary worker every year, and he was talking to me beforehand about how they bring the young boys down to join in for the work. His son was six at the time, and his son's friend, who also attended, was seven. He told me, "It's time for these boys to learn what it means to be a man. My boy ain't gonna be soft. He ain't gonna turn into no girl. And he ain't gonna be no queer."

In talking to Dave, I learned that including young boys in the branding and castration of calves is not only an important form of instruction in ranching practices for which they will eventually be responsible, but also a kind of conditioning that teaches them to have a tough exterior and an unemotional masculinity. The desensitization he described in our conversation involves burying the emotions and cultivating a thick skin. It's a social landscape where boys become "real men," and "real men" become ranchers and cowboys, and there's no room for "softness" in this transition. The boisterous celebration around it, the model by adult men of how they're supposed to behave, and the monotony that comes from castrating and branding hundreds of cows in a day thicken these calluses.

In talking with my new partner, James, who grew up in this town, he explained that this process of toughening up in the face of animal suffering was essential to sustaining practices of farming across generations. In continuing the family traditions of farming, children not only have to be trained in the actual practices of raising animals for food but also must be taught to accept and not feel as keenly the violence that becomes part of the everyday life of animal agriculture. He described it not just as a cultural norm but as a necessity for emotional and psychological survival. It's too painful to feel the suffering of each calf as they're branded or each cow bellowing across the valley after her calf has been loaded into a trailer to be transported to slaughter.

I thought about this burden on those who become farmers and the comfortable distance from the realities of animal agriculture that consumers not involved in farming enjoy. I spoke with one of the seven-year-old boys who'd attended the branding and castration party for the first time that year. I asked him how it made him feel. His eyes welled up with tears, and he said quietly, "Sad—they're just babies, and they were crying."

I hugged him and agreed, "It *is* sad. I'm sorry."

Then he said, "It was kinda fun, too. I was playing with the other kids. And Dave said this will make us men. Real cowboys."

Growing up in this community, James has observed how this early and sustained conditioning also extends out from farming practices and farmed animals. Because these ways of thinking and feeling about farmed animals are so integral to the fabric of the community and so much a part of how children become adults, he explained, they often affect how people think about and treat animals in general. When you learn intimately that violence not registered as violence against certain animals (cows, chickens, pigs, etc.) is normal, that you're directly involved in that violence, and that you're not supposed to have feelings about it, other species slip more easily into that sphere of indifference. Free-living animals, like deer, elk, bears, and moose, become targets of hunting for food and sport; others, like coyotes, wolves, and cougars, are hunted as predators whose species are perceived to threaten farmed animals. Dogs and cats become easily neglected or abused or abandoned—a widespread problem in the area, making local animal shelters impossibly overburdened. Living here, I've struggled with the visibility of animals' suffering. It's not that animal suffering isn't happening all over the place—animal suffering pervades human worlds in most places if you look for it—but perhaps it's the newness of seeing it in these particular forms that feels more potent.

At the same time, this place is filled with knowledge about animals' lifeways, their relationships, and how they navigate this dangerous multispecies world that I didn't see as often living in the city. Even among those who engage in active harm against animals, like hunters, knowledge about other species is abundant and rich, such as their understanding of the seasonal habits and movements of animals, the way their families form and then disperse, and the behaviors that bring them into the scope of the hunter's rifle or evade their tracking.

Outside of encounters of violence against animals, there are also forms of sensitivity and knowledge that accumulate over a lifetime of observation that can be a guide in gentler, more attentive ways of relating to other species. James's knowledge and that of some of the other people living here, driven by an open-hearted curiosity, shape a kind of respect and reverence for other species that I hope to engender.

James and I both live up a dirt road just out of town. In the late spring and early summer, deer give birth to their fawns, and we see them foraging in the pastures and woods or crossing the road. My first summer here, when I would drive up and down the road, I'd often see a deer with her twin fawns standing in the road. Perplexed, I asked James one day, "Why are those deer always in the road? They're gonna get hit."

And he said, "No, this is how the mom teaches her babies to know that roads are dangerous. This is a safer road—slower traffic, not many cars. It's the perfect place to learn to be afraid of vehicles. When they grow up, they'll know that any road is dangerous. It'll help them survive."

His knowledge of these deer and that what we were seeing was the education involved in their upbringing struck me hard for some reason. I hadn't considered that roads were a place to be *taught* to avoid rather than a place that animals just intuitively knew to avoid or a place where they were destined to be hit. In retrospect, *of course* animals need to learn how to navigate their worlds and evade danger—just as humans do—but my ignorance and limited thinking (even as I've tried hard to cultivate attentiveness and understanding) prevented me from even considering this. Our relationship has been filled with his observations accumulated over a lifetime, igniting electric bursts of curiosity and answers to so many of my wonderings. His own knowledge has been sparked and deepened by knowing this place, living closely with other animals, and allowing his curiosity to cultivate a sensitivity and gentleness in understanding and relating to them.

Coming from the city, where encounter with such a diverse web of animal life is much more limited, I've found (with both excitement and humility) that my skills at observation, too, are incredibly limited—humility because there is so much I don't know and haven't considered, and excitement because *there's so much I don't know and haven't considered*, opening up in front of me a lifetime of new curiosities and seeking of imperfect and incomplete knowledge. Exploring the world here with James is a reminder that *place* and the people living in it shape in profound ways the expansively varied possibilities for knowing animals differently. These ways of knowing, on the one hand, are constrained by the violence enacted in farming or hunting, for instance, and, on the other, open up limitless pathways of wonder and sensitivity that become possible in this quiet and gentle observation over a lifetime.

On our outings, my dad and I always made sure to bring along snacks—apples, cantaloupe, watermelon, carrots, that sort of thing. I got it into my head one day that the cows in the woods might like to have some of these treats as they were foraging in the dry summer heat. So, when we passed some cows on the road, I pulled off up ahead, and while my dad waited in the car, I walked back, carefully and quietly, to where the cows were standing. They were a small herd of two mothers with calves, a young steer, and another cow. They watched me warily. I stopped and spoke softly to them. I took a few steps closer, showing them the apples I held in my hands. They shuffled their hooves and continued to watch me. I took another step forward. All together, they backed away, turned, and, trotted off the road into the woods. It struck me that, here in the woods, the cows had the space and relative freedom to exercise their agency and self-determination. They weren't confined to a pen where they were forced into an encounter with a human. They saw a human approach and chose to turn away, and they had the space to do so. Of course, several months from now, they would have no choice in how they inhabited the woods; they would be rounded up and driven into transport trailers to be carried back to ranches or directly to slaughterhouses. But for now, they lived autonomously in their self-determined communities of care and kinship. Seeing that they wanted nothing to do with me, I tossed the apples into the woods as they stared at me from behind the dense brush.

The cows I had encountered during my research on dairy had often been outgoing, approaching *me* from within their enclosures, coming to the fence, reaching over, and sniffing at my clothing, licking my face, actively seeking out contact with me, a human. I had automatically assumed that these cows might be the same, but now it occurred to me that this was an absurd thing to expect; these were different individuals, in a different geographic context, with different experiences of contact with humans—and they were different breeds. Although all selective breeding of bovines involves to some extent attention to what the industry refers to as *temperament* to enable easier handling of the animals in farming practices, dairy breeds in particular have been bred with a focus on docility alongside high volumes of milk production precisely because dairy production involves so much direct physical contact between humans and cows. Cows have to be herded into milking parlors, attached to milking machines, and milked several times a day, and as such they must be docile enough to tolerate this constant contact. As pattrice jones notes, in the case of pigeons (and of cows, too), selective breeding "has profoundly influenced not only the bodies but also the psyches of individuals . . . while altering the course of evolution for entire species."[5]

Like cows raised for dairy, breeds raised for beef are also handled at key points throughout their lives as they are variously bred, branded, castrated, and then herded from pasture to woods, to transport trailer, to feedlot, to auction, to slaughterhouse. But with these cows in the cow-calf operations, there were long expanses of time when they were not in contact with humans. They lived for months in the woods without humans, and then what human contact they did have would likely have been frightening—all of the violence of the commodity-driven reproductive process, and then after a long period alone in the woods, humans on horseback suddenly rounding them up in a chaotic frenzy of panic and fear, rumbling down mountain roads on transport trailers to one destination after another, each perhaps more terrifying than the last.

It was no wonder they backed away and ran off as I approached. There was no reason I should be trusted by these cows, and perhaps I was not even a subject of curiosity, given the experiences they had already had with humans and their greater interest in being with their own kin. *What hubris!* I thought later, *thinking I could just hop out of my car, walk up to these cows, my heart filled with good intentions, to offer them some fruit.* As jones explains, "Animals don't care about our pretty ideas or pure intentions—what matters is what actually happens."[6] They "know what we *do*, not what we claim to be doing or think we are doing."[7] What I thought I was doing was offering these cows a sweet refreshing treat in the middle of summer—something that made *me* feel good to offer and which I hoped would make *them* feel good to receive. But what they saw me doing was perhaps very different—a stranger, a member of a species that had exerted control over their lives and bodies, now invading their space.

Many humans readily impose themselves on other animals' bodies and spaces. There are so many ways that this occurs—through killing animals for food or as pests or as research subjects, or through bulldozing their homes to the ground in the development of new-construction urban houses. But sometimes there are less overt forms of imposition. Sometimes it might be the result of a lack of attention, something we do without noticing we're doing it at all—for instance, tromping loudly through the woods, not realizing or thinking about the fact that this is the *actual home* of so many other species, not merely a recreational destination. Other times, it might be thoughtlessly invading the personal space or bodily autonomy of an individual—such as by touching someone in an enclosed space where they do not have the option to not be touched or by picking up someone who is smaller and more physically powerless and snuggling them to show them affection without considering whether this is something they enjoy or wish for in that moment.

I wonder what this instinct is all about. It's certainly something I've felt keenly in my life. I think part of it is about the desire for connection with members of other species. As social animals, we crave connection with other species. But so often we get it wrong. We keep wild animal species in the captivity of zoos so that we can feel connected to them in our viewing of them behind glass—the thrill of seeing a giraffe or tiger or elephant up close. Or we seek out relationships with horses through breaking, training, and riding them. Or we make claims about forms of connection with farmed animal species through farming, killing, and eating them. Efforts at connection can come in the form of touching animals, looking at them, being in close proximity to them, keeping them in cages and crates and tiny backyards, smelling and/or tasting them, killing them, or using them for recreation, but many of these forms of connection are superficial, based on fleeting encounters or instrumental relationships. Often, I think if we can sense the superficiality of these connections, like the hollowness I felt seeing orcas on the whale-watching tour, and if we are attuned to noticing the nature of these encounters, it is possible to see that even what we are craving in connection is not fulfilled. In casual conversation, I've heard so many people, for instance, in talking about visiting the zoo, say that it is exhilarating to see an animal they would otherwise not have the opportunity to see in person. But in the next breath, they reflect on their feelings of sadness at seeing the impoverished life that the animal leads. And still, they go to the zoo because there is some pull, some desire for connection, that will never in those conditions be fulfilled.

I try to take care to avoid imposing myself on other animals. And yet, still, I find myself doing it, without thinking, without even a moment of pause. With these cows, I was so overcome with wanting to do something nice for them, whose futures I knew would culminate in pain and sorrow, that I hadn't stopped to think that my very presence *as a human* might just be part of that same landscape. That being offered a piece of fruit by a human perhaps might not be a treat at all or that being approached by a human was anything but comforting or kind. My desire for connection was at least in part rooted, too, in the novelty of seeing cows in the woods and wanting to connect in some way to these individuals I didn't expect to see. It was also that I had a preconceived idea of farmed animals as animals who wanted to be touched and interacted with, and at the very least were so accustomed to it that they didn't mind. And, quite simply, I wanted to pet them, to scratch behind their ears and feel their soft noses and warm breath against my skin.

Ultimately, the connection I was seeking—and I think the connection that so many other humans seek with animals—was in large part about *me*. It wasn't about taking the time to think with careful consideration what *their* perspective might be outside of all the things that humans project on them. These are projections that should be familiar because we hear them all the time. *About a dog kept in a cage while humans are at work: "Oh, he loves his 'nest' or 'crate'—he feels safe there." Or, about a horse: "She loves to be ridden. It's one of her favorite things." Or, about cows on a dairy farm: "They want to be milked."*

What might true connection with members of other species look and feel like? Can it be that sometimes we ought not to seek connection at all?

I'm reminded of a time, years ago, when I participated in an animal studies workshop where a handful of scholars spent a few days together to offer feedback on each other's work and discuss shared interests. The conversations at the workshop were rich and productive, the opportunity to receive feedback from others working on similar issues a true gift. But we also had a curious experience one evening—made even more curious by the fact that these were scholars dedicated to thinking critically about human-animal relations. The organizer of the workshop had a long-standing relationship with the local zoo and had arranged for the group an after-hours behind-the-scenes tour of the rhino exhibit. This was a rare opportunity, and it was clear that the organizer had put a great deal of effort into coordinating this for us, drawing on the connections developed from years of critical scholarship on zoos.

First, we met the rhino keeper, who explained that the zoo had one of the few breeding programs to successfully breed a rhino and have the baby rhino survive in captivity. He explained the process of artificial insemination in which the rhino was restrained in a strong steel holding pen that held her in place to prevent resistance, and said that he had been the one to insert the semen straw into her uterus. He had the help of another zookeeper, a woman, and they referred to themselves as the baby rhino's mom and dad—a designation I found odd given the mother rhino was there, caring for her baby, and the father, long dead, was still present as a ghost in the semen that had been frozen in wait of just this opportunity to reanimate his genetic material and, in a way, bring him back to life.

We were shown a video of the rhino giving birth, a long ordeal in which the rhino paced in a small concrete room, becoming increasingly agitated as her labor progressed. We didn't have hours to spend at the zoo, so the rhino keeper kept fast-forwarding, narrating what was happening as the video sped ahead and pausing once the baby rhino started to appear. This was an amazing thing to watch. I had never seen a rhino give birth before, nor had I experienced such a clear and detailed explanation of zoos' breeding programs. It was fascinating. But I also felt uncomfortable sitting there, watching what seemed to me to an intensely private moment—the birth of this rhino's baby. I also couldn't help but think about the concrete cell into which this baby was born and the lifetime of captivity in the zoo that he would experience if he survived past his first few years. When the rhino finally finished giving birth, and the much-anticipated baby born into the world, the rhino keeper turned off the television screen and asked us, "Would you like to meet them?" Naturally, we all said "yes." Who would decline the opportunity to meet a rhino?

He led us through some back hallways and into the rear of the rhino exhibit (the part the public can't see), which was a series of three small concrete cells, just big enough for the rhinos to pace and turn in circles. In the first there was a single adult male rhino—*not the father*, the rhino keeper told us. The middle cell was empty, and the third contained the mother and her tiny baby—well, tiny for a rhino. As we crowded into the walkway in front of the cages, the mother stepped in front of her baby to shield him and retreated several steps away from the front of the cell.

"She's just protective of her baby," the rhino keeper explained. He led us back to the first enclosure with the adult male and asked us if we'd like to touch him. There were enthusiastic nods from the group. One by one, each person stepped forward to stroke the rhino's leathery skin through the steel bars of the enclosure. I don't know what it felt like because I didn't touch him. Somehow it felt wrong to touch him in a place where he didn't have the option of choosing not to be touched. I hung back, against the wall, and watched.

When I returned home after the trip, I was relaying this experience to Eric, and he asked, incredulous, "Wait, what? You didn't touch the rhino?"

"No," I said.

"Why didn't you touch the rhino? When are you ever going to get the opportunity to touch a rhino?" And he was right. That was probably the only time in my life I would have that opportunity. Because I declined, I will most likely never know what it feels like to touch a rhino.

"I don't know," I answered. "I just felt like maybe humans shouldn't be touching rhinos at all."

The experience of meeting the rhinos came back to me when I was thinking about the lives lived freely by the cows in the woods. The cows were in an unusual position to escape my company, to express their agency in not wanting to be touched. The rhinos were not in that same position. Maybe the rhino didn't mind being touched (from what I could tell, he didn't seem to mind). But in such an enclosed space, there wasn't room for him to fully express his agency, to retreat across the savannah that might have been his home under different circumstances and disappear from our presence.

Sue Donaldson and Will Kymlicka tell us that it's important to ask what kinds of relationships other animals want to have with us.[8] Some cows may want relationships with particular humans, and others may not. Similarly, pattrice jones asks us to consider what animals see and understand about us when they look at or encounter us. In writing on the Great Emu War—a military operation in the 1930s in Australia intended to wipe out large populations of emus—she asks: "Do emus see human beings as a subset of animals who have gone wrong? . . . perhaps we could think about what being allies of animals who see us as the problem really might mean. To do that, we will need to consider what emus may have learned about humans in the course of sixty thousand years of eying us warily."[9] We can ask the same of cows. What would it mean to act in solidarity with those cows in the woods, whose histories across millennia have been characterized by human control and harm? Perhaps solidarity might mean working actively to end animal agriculture and the other systems of use and harm to which cows and other species are subjected, and maybe it means that when encountering cows in the woods, we leave them alone.

As I started the car to continue our drive after my failed attempt at sharing fruit with the cows, we saw a large black cow with a tiny black newborn calf peering out from the woods on the other side of the road. The calf couldn't have been more than a week or two old. Unlike every other cow I had seen, he had no ear tag. He had been born in the woods and had not yet been found or noticed by the rancher who owned the herd. He started to nurse from his mother—hungrily suckling at her udder, a trickle of milk running down his neck. What did he then know of humans? What kinds of ingrained knowledge was he born with, and what would he learn in the coming months and year or two until his death?

Although it is often not our default position as a species that sees ourselves at the apex of evolution, we might try harder to be modest and humble in what we claim we can know about other animals—and to use this humility to honor the knowledge and wisdom of multispecies others. I wondered over the kinds of knowledges passed down through generations of cows in these woods—about the best watering spots, about the safe areas for sleeping, about the tastiest foraging places, about what it means to be bovine kin, about what these woods mean in the arc of their lives appropriated for human use, and about humans. I wonder, too, about what kinds of trauma haunt these woods, reverberating through histories of herds grazing beneath the tall trees. As jones explains, "Humans can not only traumatize individual animals but also, over time, their cultures."[10] As I explained earlier in the book, this is beginning to be acknowledged in scientific studies in the case of certain species, like orcas. Elephants' trauma, too, is being better understood—trauma that comes as a result of poaching, habitat loss, and other human impacts, and kinship networks in elephant herds are frequently shattered. In elephant herds, "intergenerational trauma can be passed down through gene expression, i.e., epigenetically, folding ruptured history into the biosocial network and into the punctured meaning of home."[11] This is compounded by the severing of social knowledge passed down through generations—in the case of elephants, this is an archive of generations upon generations of elephant knowledge transferred matrilineally, as with orcas.[12]

In cows, agricultural animal science in dairy settings has reported the trauma involved in cow-calf separation, which has multiple industry implications for the productivity of the cow and the immediate and future well-being of the calf. The resulting recommendation is to not keep cows and calves together but rather to remove calves immediately after birth on dairy farms to sever the development of a stronger bond and thus reduce the emotional and psychological (and attendant physical) effects on cows and their calves. The research on cow-calf separation is a rare acknowledgment within the agricultural industry of the trauma (what the industry refers to as *stress* or *behavioral response*) involved in routine practices of farming. However, importantly, the focus is not on the welfare of the animals for the sake of their experience but is instead oriented around maximizing productivity of both the cow and her calf. What effect does this have on the cow, whose life will be filled with these losses, and what impact will it have on the calf who is alienated from the kinds of knowledge he would gain from living in relation with his mother and herd?

Cows are not elephants who are not humans who are not chimpanzees who are not rats, and so we can't map their experiences onto each other across species. But perhaps it is not so hard to believe that cows, who are subjects of repeated separation and alienation from their kin across many generations, experience this as trauma, not only for individuals but for entire bovine lineages. If we acknowledge this possibility in cows, how might we think about the role and obligations of humans, who are, at root, the cause of this trauma—pain passed across millennia as cows are shaped and reshaped in the image of what humans want them to be?

What kinds of intergenerational trauma might be passed down in those woods? What kinds of deep relationships of care, love, and connection are formed away from humans during the months spent in these woods? And, perhaps most important, do we even need to know the particulars of these things in order to treat animals with respect and honor their capacity for self-determination?

The dominant trend in acknowledging animals' inner lives and relationships is to not acknowledge them at all until there is irrefutable evidence telling us to do so (and even then, there is resistance to taking this knowledge seriously). But what if we assumed the opposite—that animals have rich emotional and cognitive experiences, and that it is our failure to understand them that leads us to justify acts of violence? Could other animals' lives be radically reshaped? Some argue that to change people's hearts and minds about other animals, we need much more ethological research conducted on animals to better understand their capacities and prove that they deserve better care and consideration than they currently receive. I'd like to suggest, though, that maybe we don't need to know or understand everything about the intricate ways animals experience the world, but that we can, instead, appreciate the alterity of their being and trust that they are in the best position to make choices about how they live their lives.

This would involve a recognition of the rich inter- and intraspecies relationships that bring joy and meaning to their daily lives—relationships that perhaps don't require or will never include humans to flourish. In many ways, we are left with only the capacity to wonder about the nature of multispecies relationships and knowledges. For instance, are there repositories of knowledge communicated between these bovine communities and with free-living animals in the forest? Do multispecies forms of knowledge-making transgress human categorizations of *domesticated* and *wild*? And if those categories are done away with, what different ways of thinking and living and caring for others might be possible?

It was July when my dad and I found ourselves in the woods offering fruit to the cows, and wondering, in awe, at the multispecies social relationships unfolding under those towering tamarack trees. When I returned in August, that forest was on fire. Other forests, too, on my drive from Seattle, had been on fire—smoke billowing up from mountainsides in the distance, a hazy fog settling in the valleys. Forests up and down the West Coast were on fire, millions of acres burning for weeks on end. Just that week, my brother-in-law in Northern California had been evacuated with his family when a wildfire reached their backyard. *Here we are,* I thought as I drove, *living the apocalypse, with an even more apocalyptic future looming.* The valley was filled with smoke—like a thousand campfires, the smell clinging to our clothes, the bedding, the bath towels, the orange fur of Jinsky the tabby cat. The fire was six miles away, and I set my alarm for every hour through the night to check the evacuation alerts, our "go bags" packed, Jinsky's carrier ready next to the front door.

Residents in the evacuation zone who didn't have anywhere else to go took their animals—dogs and cats and goats and pigs and horses—to camp in a nearby gravel quarry where they hoped they would be safe. One of my dad's friends, who cared for dozens of animals at her home and had a small SUV to transport them to safety, we learned later, had made trip after trip between her home and the quarry with a single pig or two goats at a time. Finally, out of time, she let the chickens loose at her home and hoped they would survive the fire if it did, in fact, reach the house. The town opened its rodeo grounds to people with horses in the area who needed a place to house them, and local residents outside the fire zone who had room also took animals in. Fire trucks roared through town, blackened with soot, firefighters hanging on the sides, their faces smudged with ash and lined with exhaustion and something else—sadness, fear, the knowledge of the futility of trying to fight the fire that was already here and the many more to come? When the whole world is on fire, how do we find ways to hope, to believe that the fire can be extinguished, to take one step after another toward an unknown future?

As the forest burned, animals fled down from the wooded hills into the dry valleys. The white-tailed deer and the mule deer, the moose, the wild turkeys, the bighorn sheep, the saw-whet owls and the great gray owls, the black bears, the elk, the coyotes, the elusive lynxes and bobcats, the yellow-bellied marmots, the mountain lions, the mink and porcupines, the yellow-pine chipmunks and striped skunks, the badgers and wolverines and fishers, and the silver-haired bats. These and so many other species native to the region would have fled. Wildlife sightings proliferated as the animals were pushed down to the roads and onto residential land.

Ranchers rounded up the cows much earlier than usual, transport trailer after transport trailer carrying them away from the wildfires to their gloomy futures. I thought of that little black calf without an ear tag, and I hoped, wildly, that, somehow in the chaos of the rushed roundup as the fires spread, he and his mother had slipped away to a different forest not on fire—that they found an enchanted forest, left alone to live apart from humans, the promise of a different future.

Coda

A THUNDER OF WINGS

On a spring day, two years after I moved to this tiny town of around a hundred year-round residents in north-central Washington, I was kneeling in James's pasture, planting potatoes in long mounds. James was in an adjacent pasture working on tilling up an area for a plot of basil from which we'd make and freeze pesto. Since I'd moved here onto fifteen acres with a well but no other infrastructure, I'd been carving out a living space on a steep, meadowy hill. James lives up the road on a mostly forested forty acres where he grew up. He'd been living in the western part of Washington for ten years and had recently returned to build a small farm and live a quieter life on the land.

Over the past couple of years, we've worked together on digging out garden space at both our places for growing vegetables, fruits, and herbs. We share the labor of planting and harvesting food and putting it up for winter. We grow things like potatoes and raspberries at James's since he has more space, and onions and tomatoes at my place since I have more direct sun. We're learning the rhythm of the short, northern growing season, what does well in which spots, what plants the deer will demolish (everything), what won't grow here (sweet potatoes, for one, despite my best efforts), and how to store produce so it doesn't rot. It's a salt-of-the-earth kind of life—one shaped by the natural world around us, by the seasons, and by the animals with whom we share the environment. More than in the city, it feels like we're living in other animals' worlds. This is the world of deer, elk, moose, ground marmots, badgers, chipmunks, magpies, cougars, bobcats, grouse, and many others whose ever-presence is something we negotiate and admire.

My hands filled with seed potatoes, I looked over at James and noticed a flash of movement under the wheelbarrow next to him. It was a ruffed grouse waiting to catch worms in the newly churned soil. When she spotted one, she'd dart out and gobble them up, retreating back to the safety of the wheelbarrow to watch and wait again. She sat under the wheelbarrow for hours, preening her feathers, occasionally tucking her head under her wing to rest, but most of all she watched James.

Ruffed grouse are wild birds between the size of a large crow and a chicken. Their feathers have dark bars and spots in brownish-grays, black, and white. They have fanned tails, like turkeys, that males display during mating rituals and that females will fan out, for instance, if they are surprised and protecting their young from an intruder. Usually, though, their fans are folded in, making them look remarkably plain, blending easily into their habitat. They tend to live in dense undergrowth with a view of their surroundings, vigilant in watching for predators.

In the spring, grouse are solitary birds, joining into flocks in the fall after the females have laid and hatched their eggs and raised their young during the summer. They mostly stay on the ground, moving under cover, but when startled will suddenly burst into flight, their wings making a loud wave of sound.

Grouse are considered "game" birds and are hunted for meat when they come together in flocks between September and January each year. Hunters flush them out of their hiding spots, and they fly up, all together—a flock on the wing. The unlucky ones are shot out of the air, gathered up, plucked, and prepared for the dinner table.

It all began when this particular female grouse started showing up at James's, perching on the roof of the house. She would fly down and peck around the grass, nibbling on weeds and the occasional insect, and then retreat to the roof again when the dogs were loose in the yard and she sensed danger. A few days after she showed up for the first time, she started following James around, curious about what he was doing, and clucking at him while he worked. He would sit in the grass, and she would walk up and crouch down, nestling against him, her wings extended out at her sides waiting for him to pet her. She'd sit there like that, making a soft cooing sound as he stroked her silky feathers and rubbed the base of her neck. When he went back to work, she would wander around the yard, checking in on him frequently. She was never far away.

Part of what was remarkable about Winnie's choice to spend her time with James was that he never fed her. Often, we build relationships with free-living animals through feeding them or giving them occasional treats, as I'd done with the crows in Seattle and as I've done here in my new home with a community of chipmunks who come to sit on my knee or in my hand, looking up at my face with curiosity and trust. Winnie was different. She chose a relationship with James not cultivated through being fed. The trust she felt for him was about the two of them getting to know each other, finding joy in the connection, and building an intimacy that somehow seemed purer than those relationships we form with animals we feed.

It was this trust, this curiosity, and their shared desire for knowing one another that defined their connection. And after she'd been showing up for a couple of weeks, he decided to give her a name, ruminating over the perfect one, and settling on "Winnie."

One evening, James was late coming over for dinner. I had just begun to wonder where he was (he'd texted half an hour earlier to let me know he was on his way) when I saw his truck coming slowly—unusually slowly— up my driveway. As he reached the top, he leaned out the window and called, "Look!" He pointed back at the covered bed of the pickup, and I saw Winnie sitting there. When the truck stopped, she walked to the edge of the truck bed and hopped down into the driveway. We sat down in the rocky dirt near the truck and watched her. She pecked around, eating some of the weeds, periodically coming over to James and crouching still so he could pet her.

James explained that he'd been heading out of his driveway when he saw Winnie barreling up the drive after him. He stopped at the top, and she flew up onto the gate and then onto the roof of the parcel box at the entrance to the driveway. From there, she hopped onto the truck bed and settled in. He'd climbed out of the truck and offered her his arm to lift her back down to the ground, but she refused. He waited, thinking she'd hop down, but she didn't budge. Finally, he said to her, "OK, you're just going to come with me then." He started driving, and she hopped through the open back window of the cab onto the top of the back seat for the journey, climbing back out onto the bed when they made it to my driveway.

We sat with her for an hour until it started to get dark, and we realized she was far from the safety of wherever it was on James's land she slept at night. As if she sensed this at the same time, she suddenly took off into the tall grass and up the hill in the opposite direction from James's place. *Where would she sleep?* we worried. Ruffed grouse typically have a home range of ten to forty acres, and this was outside that distance. It was unfamiliar territory, and there was a coyote thoroughfare on the hill near where she had disappeared.

"What if she gets eaten by a coyote?" I worried. "We can go get her. She's still close. We could take her back to your place."

"No," he stated and then paused. "She's a wild animal. She can take care of herself."

"OK," I said. And then, "But what if she gets eaten by a coyote?"

"She might. But I think she'll be fine," James said.

"We could still go get her," I suggested, stuck in a loop of worry.

"No, she's a wild animal," James concluded confidently.

We went inside, worrying about her fate, both of us questioning whether we were going to see her again. But the next day, James was working in his pasture, and Winnie came walking down the driveway, joining him for a day of chores and crouching down periodically for some petting, hopping onto his arm when he went to stand up.

That night, he was getting ready to head over for dinner again. He hadn't seen Winnie for the past few hours and figured she was off doing grouse things. But hearing his truck engine start, she came flying up from the creek where she had disappeared for the afternoon. She landed on the back of the truck and again refused to get off. She rode with him up to my place, again hanging out with us in the driveway until dusk fell, and again disappearing up the hill for the night. The next day, Winnie didn't show up, and we were sure she'd been caught in the night by a coyote or bobcat. But the following day, James was leaving for an outing to town, and he saw her walking up the middle of the road toward his place, intrepid in her determination to return home.

It went on like this for about a week. She'd ride with him in the evenings, spend the night up the hill, and return to spend the day with him while he worked. She became more and more affectionate, demanding to be petted, cooing softly, nestling against him, never letting him far from her sight. She had chosen him.

It was around this time we started hearing male grouse making their mating call. During mating season, the males beat their wings in a display that creates a sonic boom, resulting in a drumming sound. This sound echoed out from the woods, where they would be standing on a log or stump, their tails fanned out, trying to attract a female. Winnie seemed to ignore this for a while, preferring to stay with James as she foraged for food and sought attention.

Then, one day, she disappeared. We imagined that she was out in the woods somewhere making her nest in the undergrowth, creating a soft depression in a safe spot, and lining it with leaves and feathers to keep her eggs warm. We imagined her laying her cream-colored eggs—between eight and twelve based on what we'd learned researching grouse. We imagined her sitting on them diligently for several weeks until they hatched, taking only short breaks each day to find food for herself. We imagined her caring for her brood of chicks while they learned to find food and fly and departing soon after, leaving them to find their way in the world on their own. We expected her to emerge from the woods any day with a healthy flock, introducing them to James and returning to her routine of spending the days with him, chicks in tow. We imagined her running up the driveway after his truck with her chicks, and the flock flying up onto the truck bed to come here for the evening. But we didn't see her. She didn't appear as we'd hoped.

One day, we were walking in the dense woods at James's and suddenly almost stepped on a grouse and her brood. Chicks scattered into the brush, and their mother stood her ground, spreading her wings and fanning her tail, hissing loudly and lunging toward us. We were as startled as she was. We stumbled back as she came at us aggressively, protecting her young, and retreated to a different path. *Was it Winnie?* we wondered. *No,* James decided, *she doesn't look like Winnie.*

Over the coming weeks, we saw flocks of young grouse foraging for food around the area, and we wondered which one was Winnie's. Ruffed grouse have always made their home at James's, where they have ample forest cover to hide and move around with relative safety. But we'd never seen grouse (other than Winnie when she came visiting) at my place—the meadow ecosystem being prime hunting ground for birds of prey who can easily spot small animals moving through the grass. But that summer, a flock of grouse made their home here on the hillside. I see them poking around the driveway, eating weeds, preening their feathers, clucking and cooing at each other. I see the meadow grass rustle and now recognize when that rustling is a flock of grouse. I see them drinking the crisp, clean water in the natural spring behind the house, gathering in the open area around it, always on alert.

Sometimes when I walk out my front door, not knowing they're there, I startle them into flight, alarming all of us in the unexpected encounter. Every time, I wonder if Winnie is one of them, having brought her brood here to live—or imagining that they go back and forth to James's, walking up and down the road, making both our places their home.

James and I still wait to see Winnie again as fall has turned to winter and the ground is covered in snow. When we see a female grouse, we look discerningly to see if we recognize her. Or, rather, *James* observes them discerningly, and I look to him to see any recognition. I always defer to him in deciding whether it's Winnie, my identification capacities being still as woefully inept as they were years before with the crows. We wait and watch, hopefully, determined in the belief that she'll come back and wondering if, perhaps, she'll appear to James again next spring.

But it's more likely we won't see her again, and that's the nature of living in shared spaces with animals. We pass through each other's lives. We might just glimpse each other from a distance as we move through our days, or we might form intense, intimate bonds for a short time before our paths part. There's beauty and mystery in these encounters—in wondering what kind of encounter it will be.

What felt perfect and true about Winnie's presence was that it was wholly on her own terms. It was she who decided one day to walk out of the woods and introduce herself to James. It was she who decided to spend her days following him around and nestling against him, cooing when he sat down to pet her. It was she who hopped into his truck and rode with him up to my place. It was she who walked back up the road each day to return to him. And it was she who left when there were other things for her to attend to. There's something about her agency and his honoring of it—the way *she* determined the relationship she wanted to have—that I think offers a path forward for how we might relate differently to the animals around us.

How can we cultivate this sensitivity, this openness, this acceptance of animals on their own terms and the choices they make in what sorts of relationships they want to have with us? Winnie taught us that we can practice attentiveness and care through honoring and respecting animals' self-determination and their secret lives that are all their own. Sometimes animals choose us and invite us into their worlds. There's magic in that feeling of being chosen to glimpse even fleetingly the intimate lives of others. But sometimes it's our absence that animals choose. The cows I saw in the woods chose *not* to have an encounter with me, retreating into the trees when I approached. The deer in the driveway who scatter when I emerge from the house choose the company of other deer. The coyotes who stand majestically on the hill above my home trot away when they spot me watching them, turning back to confirm that I'm not pursuing them.

These fleeting moments have taught me that often we ought not seek connection with animals in whose worlds we live at all. There's beauty and tenderness in accepting the things we don't and cannot know, in letting their lives be a mystery, and in recognizing that this is how they might flourish most. And perhaps there's a deeper connection that comes from that—in appreciating from a distance the kinds of lives they want to live, noticing and not acting on our desires to be close to them, and ultimately letting them be.

Years after meeting the crows, and in the fresh wake of fleetingly knowing Winnie, I still listen for the sound of feathers. The sound is a sigh, a rustle of crow feathers in the wind sometimes so quiet I think I might have imagined it, other times a thunder of grouse wings impossible to ignore. It is a haunting, a trace, a whisper, and a roar of what was and what could be. I think of these sounds of feathers—of wings in flight—as a reminder of all that haunts us, of the histories of violence and pain and alienation that deliver us to where we are now. They're a recollection of desire and belonging, kinship and care, a longing for pasts we never had and futures we haven't yet imagined. They're a quiet call and a resounding demand for the most delicate forms of attentiveness. And in this attentiveness— in this nearly silent whisper or wave of urgent sound—there is a flutter of hope, of anticipation, of the possibilities for repair, of futures lost and gentler futures found.

Notes

Introduction. Listening for the Sound of Feathers

1 Marzluff and Angell, *In the Company of Crows*; Marzluff and Angell, *Gifts of the Crow*.

2 Odell, *How to Do Nothing*, 81.

3 Horowitz, *On Looking*, 263.

1. Knowing Our Haunted Homes

1 Marx, *Capital*.

2 See Moreton-Robinson, *White Possessive*.

3 For discussions of how anthropocentrism might be defined and understood, see Calarco, "Being Toward Meat"; Collard, *Animal Traffic*; Probyn-Rapsey, "Anthropocentrism"; Srinivasan and Kasturirangan, "Political Ecology."

4 TallBear, "Indigenous Reflection," 234. See also Goldberg-Hiller and Silva, "Sharks and Pigs"; Simpson, *As We Have Always Done*.

5 See Cacho, *Social Death*.

6 Much attention has been given to constructions of race, for instance, in the United States in the binaries between white people and Black,

Indigenous, Asian, and Latine at different historical moments and in sustained ways that track into the present. But it's not just white identity that creates the subhuman or nonhuman others. Nonwhite racial groups also construct these hierarchies between non-white racial categories (e.g., Latine or Asian racism toward Black or Indigenous people) or even within a single race (e.g., the privileging of lighter-skinned versus darker-skinned Black people). Racial hierarchies become even more complicated in a global context, where religious and cultural factors work to dehumanize certain groups (e.g., in India, where the caste system is a complex hierarchy drawing from many factors that organize Indian society and where Hinduism is leveraged to demonize Indian Muslims, or in Peru, where those of Spanish descent discriminate against Indigenous Peruvians, like the Quechua peoples). But race is not, by far, the only mechanism for dehumanizing certain human groups. People with disabilities have long been framed as animalistic, with their cognitive and intellectual or physical characteristics being framed as "abnormal" and thus inferior—to be eradicated through practices like eugenics. Geography can also be used as a weapon of dehumanization, such as framing rural folks as "backward"—those living lives against the grain of urban cultures, often in poverty (another hierarchizing site of dehumanization), such as communities in remote places in Appalachia. Gender persists as a site of inferiority, evidenced, for instance, in the persistent failure of the Equal Rights Amendment, the wage gap, and the assault on women's reproductive rights in the United States. The list goes on—there is no shortage of hierarchical relations within the human species that work to define a highly valued, narrowly defined conception of the full human. There are many excellent texts that provide comprehensive discussions of constructions of the *human* and *nonhuman* in a hierarchical ordering as the basis for othering. See, for instance, Boisseron, *Afro-Dog*; Deckha, "Subhuman as a Cultural Agent"; Glick, "Animal Instincts"; Jackson, "Animal"; Jackson, *Becoming Human*; Kim, *Dangerous Crossings*; Ko, "Addressing Racism"; Narayanan, "Cow Protection"; Narayanan, "Animating Caste."

7   McKinley and Smith, *Handbook of Indigenous Education*; see also Wolfe, "Settler Colonialism."

8   Odell, *How to Do Nothing*, 19.

9   Gordon, *Ghostly Matters*, 63–64.

10 Tuck and Ree, "Glossary of Haunting," 642.

11 Gordon, *Ghostly Matters*, 63.

12 Gordon, *Ghostly Matters*, 64.

13 Rose, "Dialogue," 129.

14 Quoted in Klein, "Dancing the World into Being."

15 See de la Cadena, "Invitation to Live Together."

16 Albrecht et al., "Solastalgia," 96.

17 Center for Whale Research, "Southern Resident Orcas and Salmon."

18 Center for Whale Research, "Southern Resident Orcas and Salmon."

19 Mapes, "Southern Resident Orca Matriarch."

20 KING Staff, "3 More Southern Resident Orcas"; Mapes, "Southern Resident Orca Matriarch."

21 Mountain, "How the First Orcas Were Captured." Also see the documentary *Blackfish*, directed by Gabriela Cowperthwaite, for video documentation of this process of capture.

22 Mountain, "How the First Orcas Were Captured."

23 Marino et al., "Neuroanatomy of the Killer Whale"; Marino et al., "Cetaceans Have Complex Brains"; Orca Nation, "Social Intelligence of Orcas."

24 Mapes, "Grieving Mother Orca."

25 King and Shepard, "2 Newborn Orcas."

26 See Bekoff, "Animal Consciousness"; Bekoff, "Animal Emotions"; Burghardt, "Anthropomorphism"; Horowitz and Bekoff, "Naturalizing Anthropomorphism"; Karlsson, "Critical Anthropomorphism."

27 jones, "Stomping with the Elephants," 327.

2. Journeys Ended Here

1 Lopez and Eschner, *Apologia*, 2.

2 Forman and Alexander, "Roads and Their Major Ecological Effects," 213.

3 Lopez and Eschner, *Apologia*, 13.

4 Foer, *Eating Animals*, 37.

5 Watson, "Remains to Be Seen."

6 United States Federal Highway Administration, "America's Highways."

7 United States Federal Highway Administration, "America's Highways."

8 United States Federal Highway Administration, "America's Highways," 18.

9 Annals of Congress.

10 United States Federal Highway Administration, "America's Highways."

11 Annals of Congress.

12 Mamers, "Human-Bison Relations."

13 Chinese workers made up 90 percent of the labor force dedicated to railroad construction in the West. Conditions of desperation driven by war, foreign military intervention, and poverty in China's Pearl River Delta in the nineteenth century drove the migration of Chinese citizens to western North America for paid work on the railroads. Spotty recordkeeping obscured the role of enslaved laborers in the construction of the railroad, but they were, in fact, a significant source of forced railroad labor, especially throughout the South. Railroad companies both purchased enslaved people outright and rented them from plantation owners. Conditions working on the railroad were highly exploitative: Housing and food provision was abysmal, the work itself was extremely dangerous (especially "blasting"), and diseases spread rampantly in railroad encampments with little or no available medical care. This work was so dangerous that plantation owners were reticent to hire out the enslaved people they owned; when they did so, they often stipulated that the enslaved laborers should not be involved in the most dangerous forms of work on the railroads. In short, the labor required to build the railroads routinely put the lives and well-being of human laborers at extreme risk, disproportionately affecting society's most vulnerable people (just as the most dangerous work continues to be performed today).

14 Day, *Alien Capital*; Karuka, *Empire's Tracks*; Sharpe, *In the Wake*; Voss et al., "Archaeology of Precarious Lives."

15 Cronon, *Nature's Metropolis*, 74.

16 Kroll, "Environmental History of Roadkill."

17 United States Federal Highway Administration, "Our Nation's Highways."

18 United States Federal Highway Administration, "Our Nation's Highways."

19 Marx, *Capital*.

20 Soron, "Road Kill."

21 Freund and Martin, "Commodity That Is Eating the World."

22 Interrante, "Road to Autopia," 93.

23 Kroll, "Environmental History of Roadkill."

24 United States Federal Highway Administration, "Our Nation's Highways."

25 United States Federal Highway Administration, "Our Nation's Highways."

26 United States Federal Highway Administration, "Our Nation's Highways."

27 United States Federal Highway Administration, "Our Nation's Highways."

28 United States Federal Highway Administration, "Our Nation's Highways."

29 United States Federal Highway Administration, "Our Nation's Highways."

30 Soron, "Road Kill," 122.

31 Soron, "Road Kill," 120.

32 Soron, "Road Kill," 120.

33 Monahan, "Mourning the Mundane," 153.

34 Quoted in Simmons, *Fins, Feathers and Fur*, 36.

35 Cool Geography, "Urban Physical Waste Generation."

36 Kisiel, *At Rest*.

37 Watson, "Remains to Be Seen," 150.

38 Watson, email to author, November 7, 2019.

39 United States Forest Service, "Snake Migration."

40 United States Forest Service, "Snake Migration."

41 Kroll, "Environmental History of Roadkill."

42 I-90 Wildlife Bridges Coalition, "Project Info."

43 I-90 Wildlife Bridges Coalition, "Project Funding"; I-90 Wildlife Bridges Coalition, "A Historic Coalition."

44 Corday, "Road Stories."

45 Soron, "Road Kill," 123–24.

46 Legendre, "Degrowth Movement."

47 Tuck and Yang, "Decolonization."

48 Marshik et al., "Preserving a Spirit of Place."

3. The Scent of the Spectral

1   Washington, *Medical Apartheid*. See also Hatch, *Silent Cells*, for a discussion of some of the complexities of contemporary prison experimentation. Hatch argues, for instance, that participation in drug trials is desirable and is often limited to more privileged prisoners (e.g., white prisoners or those who have demonstrated good behavior).

2 The legislation passed after a public outcry the previous year, when a dalmatian was stolen from a family in Pennsylvania, sold to a hospital in the Bronx, and later died in an experimental medical procedure. The death of this dog, named Pepper, launched a media investigation into what turned out to be a widespread practice of animal dealers stealing dogs and selling them to research facilities. The investigation also revealed the conditions under which dogs were bred in breeding facilities that sold dogs for research.

3 The Animal Welfare Act defines *animal* as follows: "The term 'animal' means any live or dead dog, cat, monkey (nonhuman primate mammal), guinea pig, hamster, rabbit, or such other warm-blooded animal, as the Secretary may determine is being used, or is intended for use, for research, testing, experimentation, or exhibition purposes, or as a pet; but such term excludes (1) birds, rats of the genus *Rattus*, and mice of the genus *Mus*, bred for use in research, (2) horses not used for research purposes, and (3) other farm animals, such as, but not limited to livestock or poultry, used or intended for use as food or fiber, or livestock or poultry used or intended for use for improving animal nutrition, breeding, management, or production efficiency, or for improving the quality of food or fiber. With respect to a dog, the term means all dogs including those used for hunting, security, or breeding purposes." 7 U.S.C. § 2132—Definitions.

4 USDA, "Annual Report Summary," 1.

5 USDA, "Annual Report Summary," 1.

6 Carbone, "Estimating Mouse and Rat Use," 1.

7 Taylor and Alvarez, "Estimate of the Number of Animals," 205.

8 *Online Etymology Dictionary*, "Abstracted."

9 Dodman, "Canine PTSD"; Sullivan, "How to Heal."

10 Dodman, "Canine PTSD"; Mikics et al., "Rats Exposed to Traumatic Stress."

11 Horowitz, *Being a Dog*, 50.

12 Giraud and Hollin, "Care, Laboratory Beagles," 43.

13 Giraud and Hollin, "Care, Laboratory Beagles," 31.

14 Giraud and Hollin, "Care, Laboratory Beagles," 39.

15 IACUCs must consist of at least five people, including "one veterinarian with training or experience in laboratory animal science and medicine, who has direct or delegated authority and responsibility for activities involving animals at the institution; one practicing scientist experienced in research with animals; one member whose primary concerns are in a nonscientific area (e.g., ethicist, lawyer, member of the clergy); and one member who is not affiliated with the institution

other than as a member of the IACUC." National Institutes of Health Office of Laboratory Animal Welfare, "IACUC."

16   These anxieties intensified again in the lead up to the university breaking ground in 2015 on the massive new underground Animal Research and Care Facility, costing $142 million. Broom, "Hundreds of Activists."

17   Gluck, *Voracious Science*, 176.

18   Instances of noncompliance are recorded for species like nonhuman primates, dogs, and cats and can include strangulation on enrichment devices in monkeys' cages, death from not meeting basic standards of surgical practice and/or improper use of anesthetics during surgery, injuries sustained by other animals, and death from starvation and dehydration as a result of staff not conducting checks of automated waterers and feeders and/or falsifying health and wellness check records. These are all examples of documented USDA violations at the University of Washington (see University of Washington Office of Animal Welfare, "Institutional Compliance"); however, *most* "adverse events" in animal research—and most reported in the University of Washington's IACUC meetings, specifically—will not appear in APHIS inspection reports because they occur with members of species not covered by the Animal Welfare Act and as such are not subject to USDA inspections.

19   Gluck, *Voracious Science*, 184.

20   As far as the public can tell, at the time of writing this book, the University of Washington does not have an ethicist, or anyone trained in careful ethical deliberation, on its IACUC. The public is not permitted to know the names of the IACUC members, only initials and a few words describing their qualifications. These qualifications reflect that the committee is nearly completely composed of animal researchers, laboratory veterinarians, and others with a vested interest in sustaining the status quo of animal research.

21   Gluck, *Voracious Science*, xii.

22   Gluck, *Voracious Science*, x, 282

23   This logic reflects the "three R's" of animal research, a system designed to encourage researchers to think more carefully about their use of animals. The three R's are *reduction*, *refinement*, and *replacement*. *Reduction* is focused on reducing the number of animals who are used in research overall—and this can occur in several ways, such as using an animal as their own control to avoid the needless replication of existing studies. *Refinement* calls for methods that limit as much as possible the pain and trauma an animal experiences in research. *Replacement* refers to both the development of technologies that replace the use of animals entirely and the use of what are

understood as "lower-order" species (e.g., mice or rats) in the place of "higher-order" species (e.g., dogs or primates).

24 Bekoff, "Sentient Rats."

25 Heath, "Animals in Research."

26 Heath, "Animals in Research."

27 Kalueff et al., "Time to Recognize Zebrafish," 1023.

28 Perathoner et al., "Potential of Zebrafish," 456.

29 Taylor, "Precarious Lives of Animals," 62.

30 We know, though, that even being human doesn't prevent unethical use and exploitation as historical and contemporary examples of often-racialized human exploitation for medical research illustrate.

31 Foundation for Biomedical Research, "New Billboards"; Sharp, *Animal Ethos*.

32 Foundation for Biomedical Research, "New Billboards"; Sharp, *Animal Ethos*.

33 Foundation for Biomedical Research, "Animals Behind Top Drugs."

34 Foundation for Biomedical Research, "Animals Behind Top Drugs."

35 Government Accountability Office, "National Institutes of Health."

36 Knight, "Animals in Research."

37 Herrmann, "Refinement," 19; see also Knight, "Animals in Research."

38 Herrmann, "Refinement," 19; see also Scherer et al., "Full Publication."

39 Herrmann, "Refinement," 19; see also Ioannidis, "Why Most Published Research Findings Are False."

40 Government Accountability Office, "National Institutes of Health."

41 Garner, "Significance of Meaning," 440.

42 Garner, "Significance of Meaning," 440.

43 McManus, "Ex-Director Zerhouni."

44 Physicians Committee for Responsible Medicine, "Asthma Research"; see also Corry and Irvin, "Promise and Pitfalls."

45 Benam et al., "Small Airway-on-a-Chip."

46 Rozenbaum, "Could AI Replace Animal Research?"

4. Consumed by Desire

1 Roeder, "How to Sex Baby Chicks."

2 Roeder, "How to Sex Baby Chicks."

3 *Merriam-Webster Dictionary*, "Consume."

4   The only federal law governing the welfare of farmed animals is the Humane Methods of Livestock Slaughter Act, which sets standards for what the government determines is involved in "humane slaughter." The Animal Welfare Act, which offers more widespread welfare protections, excludes farmed animals entirely. The Humane Methods of Livestock Slaughter Act excludes all birds, fish, and rabbits, who have no welfare protections under federal law. They typically do not have welfare protections in state law either, as many states' anticruelty laws contain exemptions for farmed animals. Rabbits likely get lumped into the "poultry" category at auction, mirroring the legal categorization of protected/unprotected species in farming.

5   Hirsch, "Detailed Discussions of Legal Protections."

6   Faunalytics, "Global Animal Slaughter."

7   Fredrickson, "Ovarian Tumors of the Hen."

8   Fredrickson, "Ovarian Tumors of the Hen"; Johnson, "Standard of Perfection."

9   Giles et al., "Restricted Ovulator Chicken"; Johnson, "Standard of Perfection."

10   Johnson, "Standard of Perfection."

11   United States Department of Agriculture, "Milk Production," 2.

12   Hubbard, "Buffalo Genocide," 292.

13   Hubbard, "Buffalo Genocide"; Justice, "Relevant Resonance."

14   Hubbard, "Buffalo Genocide."

15   Hubbard, "Buffalo Genocide," 299.

16   Hubbard, "Buffalo Genocide," 299; Isenberg, *Destruction of the Bison.*

17   Hubbard, "Buffalo Genocide," 297.

18   Anderson, *Creatures of Empire.*

19   Wolfe, "Settler Colonialism," 388.

20   Hubbard, "Buffalo Genocide," 298.

21   Hubbard, "Buffalo Genocide," 295.

22   Netz, *Barbed Wire.*

23   More recently there has been an increase in the use of hormone implants that prevent egg laying in chickens as an alternative to feeding eggs back to the birds. For more on the topic of implants in rescue contexts that use them, see Narayanan, "Ecofeminist Politics."

24   Focusing on the settler colonial context of farming animals, the orienting relationship is one of extraction. The crafting of farmed animal species as colonial subjects, fashioned, controlled, and materially and

conceptually framed as "resource," is tied to their weaponization in the settler colonial and capitalist projects of land dispossession and occupation, environmental destruction, and other attendant harms. At the heart of these extractive relations are the animals themselves—the Emilys, Charlottes, Janes, and Georges—whose experiences are characterized by humans stealing their lives from them, both in their killing and in their exploitation for reproductive outputs, like the eggs they lay.

For many Indigenous communities, eating animals is framed as a relationship of reciprocity (especially in relationship to practices of hunting). For the Michi Saagiig Nishnaabeg scholar Leanne Betasamosake Simpson, in *As We Have Always Done*, "the alternative to extractivism is deep reciprocity. It's respect, it's relationships, it's responsibility, and it's local" (75). For a discussion of this topic, see also Nadasdy, "Gift in the Animal." My focus here is not on Indigenous practices of eating animals but rather on the settler and capitalist logics that govern the consumption of farmed animals.

Our relationship of care for the chickens and our consumption of their eggs as one part of this care could be described as a reciprocal relationship—not as transactional but as one of mutual care and flourishing. It would be easy to focus on our co-flourishing so long as the chickens were well and weren't showing the harms of such intensified egg production. And yet, egg production takes its toll in real and catastrophic ways on hens' bodies, as does all milk and meat production as well—a toll that I would suggest can't be repaired or ameliorated by reciprocal relations or narratives of consumption. Reciprocity here, then, may not make possible relationships of co-flourishing when the very nature of hens' (and other animals') beingness has been crafted and cultivated by human priorities for their productive capacities, and as the continued consumption of their reproductive outputs sustains conditions of extraction and harm in their bodies and lives. Humans' consumption of animals both in the form of reproductive outputs like eggs or milk and in the form of their bodies as meat takes much more from other animals than we give. The stakes of reciprocity where eating animals is concerned are much higher for other animals—their very existence is at stake, weighed against the pleasure and desire (and sometimes need) of humans for a meal.

5. Ghosts in the Garden

1 Gan et al., "Haunted Landscapes," G6.

2 Tuck and Yang, "Decolonization," 5.

3 Brown, *City Is More Than Human*, 17.

4   Brown, *City Is More Than Human*, 17.

5   Brown, *City Is More Than Human*, 21.

6   Thrush, *Native Seattle*, 94.

7   Thrush, *Native Seattle*, 95.

8   Brown, *City Is More Than Human*, 159, 76, 215.

9   Brown, *City Is More Than Human*, 220.

10 Brown, *City Is More Than Human*, 136.

11 Kaukeinen, "Myths About Rodents."

12 Hubbard and Brooks, "Animals and Urban Gentrification," 1495. See also German and Latkin, "Exposure to Urban Rats"; Lipstein, "You're Not Mapping Rats."

13 Hubbard and Brooks, "Animals and Urban Gentrification," 1495.

14 Hubbard and Brooks, "Animals and Urban Gentrification," 1495.

15 King County Public Health, "Diseases from Rodents."

16 King County Public Health, "Diseases from Rodents."

17 Wiesman et al., "Tularemia in Washington."

18 Washington State Department of Health, "Leptospirosis."

19 King County Health Department, "Canine Leptospirosis."

20 Shanahan, "New York City Rats."

21 Dean et al., "Human Ectoparasites."

22 Humane Society of the United States, "What to Do About Wild Rats."

23 *Merriam-Webster Dictionary*, "Forget."

24 Hubbard and Brooks, "Animals and Urban Gentrification," 1505.

25 jones, "Stomping with the Elephants," 322.

26 Schuerman, *Newcomers*, 7.

27 In 1990, Beacon Hill was made up of 76.3 percent people of color, but this number decreased to 65.3 percent by 2018. Two of the starkest changes, in addition to the growing white population, have been a growth in the Latine population (an increase from 0.4 percent of the population in 1990 to 16.9 percent in 2018) and the dramatic decrease of the Asian and Pacific Islander population (this population, which constituted 50 percent of the neighborhood in 1990, has been cut nearly in half—to 27.7 percent in 2018). City of Seattle, "Neighborhood Change."

28 Simpson, *As We Have Always Done*, 153.

29 Gumbs, *Undrowned*, 55.

30 Shotwell, *Against Purity*, 8.

6. On Cows in the Woods

1   Flower and Weary, "Effects of Early Separation."

2   jones, "Property, Profit," 7–8.

3   jones, *Oxen at the Intersection*, 19.

4   Stanescu, "Selling Eden."

5   jones, "Harbingers of (Silent) Spring," 202.

6   jones, "Eros and the Mechanisms of Eco-Defense," 103.

7   jones, "Provocations from the Field," 15.

8   Donaldson and Kymlicka, *Zoopolis*, 100.

9   jones, "Provocations from the Field," 14.

10   jones, "Provocations from the Field," 6.

11   Willett, *Interspecies Ethics*, 140.

12   McComb et al., "Matriarchs as Repositories."

Bibliography

Albrecht, Glenn, Gina-Maree Sartore, Linda Connor, et al. "Solastalgia: The Distress Caused by Environmental Change." *Australian Psychiatry* 15 (2007): 95–98.

Anderson, Virginia. *Creatures of Empire: How Domestic Animals Transformed Early America*. Oxford University Press, 2004.

Annals of Congress. 14th Cong., 2nd Sess. Gales and Seaton, Washington, DC, 1854, 851–54.

Bekoff, Marc. "Animal Consciousness and Science Matter: Anthropomorphism Is Not Anti-Science." *Relations Beyond Anthropocentrism* 1, no. 1 (2013): 61–68.

Bekoff, Marc. "Animal Emotions: Exploring Passionate Natures." *BioScience* 50, no. 10 (2000): 861–70.

Bekoff, Marc. "Sentient Rats: Their Cognitive, Emotional, and Moral Lives." *Psychology Today*, February 2, 2020.

Benam, Kambez H., Remi Villenave, Carolina Lucchesi, et al. "Small Airway-on-a-Chip Enables Analysis of Human Lung Inflammation and Drug Responses in Vitro." *Nature Methods* 13 (2016): 151–57.

Boisseron, Bénédicte. *Afro-Dog: Blackness and the Animal Question*. Columbia University Press, 2018.

Broom, Jack. "Hundreds of Activists Protest Against New UW Animal Lab." *Seattle Times*, April 25, 2015.

Brown, Frederick L. *The City Is More Than Human: An Animal History of Seattle*. University of Washington Press, 2016.

Burghardt, Gordon. "Anthropomorphism: Critical Anthropomorphism." *Encyclopedia of Animal Rights and Animal Welfare* 1 (2010): 73–74.

Cacho, Lisa Marie. *Social Death: Racialized Rightlessness and the Criminalization of the Unprotected.* New York University Press, 2012.

Calarco, Matthew. "Being Toward Meat: Anthropocentrism, Indistinction, and Veganism." *Dialectical Anthropology* 38, no. 4 (2014): 415–29.

Carbone, Larry. "Estimating Mouse and Rat Use in American Laboratories by Extrapolation from Animal Welfare Act–Regulated Species." *Scientific Reports* 11, no. 493 (2021): 1–6.

Center for Whale Research. "Southern Resident Orcas and Salmon." Accessed November 19, 2021. https://www.whaleresearch.com/orcassalmon.

City of Seattle. "Neighborhood Change." Last updated April 2019. https://population-and-demographics-seattlecitygis.hub.arcgis.com/pages/neighborhood-change.

Collard, Rosemary-Claire. *Animal Traffic: Lively Capital in the Global Exotic Pet Trade.* Duke University Press, 2020.

Cool Geography. "Urban Physical Waste Generation." Accessed April 16, 2025. https://www.coolgeography.co.uk/advanced/Urban_Physical_Waste_Generation.

Corday, Jackie. "Road Stories." Animal Road Crossing. Accessed August 18, 2024. https://arc-solutions.org/road-stories/.

Corry, David B., and Charles G. Irvin. "Promise and Pitfalls in Animal-Based Asthma Research." *Immunologic Research* 35, no. 3 (2006): 279–94.

Cowperthwaite, Gabriela, dir. *Blackfish.* Magnolia Pictures, 2013.

Cronon, William. *Nature's Metropolis: Chicago and the Great West.* Norton, 1991.

Day, Iyko. *Alien Capital: Asian Racialization and the Logic of Settler Colonial Capitalism.* Duke University Press, 2016.

Dean, Katharine R., Fabienne Krauer, Lars Walløe, et al. "Human Ectoparasites and Spread of Plague in Europe." *Proceedings of the National Academy of Sciences* 115, no. 6 (2008): 1304–9.

Deckha, Maneesha. "The Subhuman as a Cultural Agent of Violence." *Journal of Critical Animal Studies* 8, no. 3 (2010): 28–51.

de la Cadena, Marisol. "An Invitation to Live Together: Making the Complex 'We.'" *Environmental Humanities* 11, no. 2 (2019): 477–88.

Dodman, Nicholas. "Canine PTSD." *Psychology Today*, October 25, 2016. https://www.psychologytoday.com/us/blog/dog-days/201610/canine-ptsd.

Donaldson, Sue, and Will Kymlicka. *Zoopolis: A Political Theory of Animal Rights.* Oxford University Press, 2011.

Faunalytics. "Global Animal Slaughter Statistics and Charts: 2020 Update." Updated July 29, 2020. https://faunalytics.org/global-animal-slaughter-statistics-and-charts-2020-update/.

Flower, Frances, and Daniel Weary. "Effects of Early Separation on the Dairy Cow and Calf: Separation at 1 Day and 2 Weeks After Birth." *Applied Animal Behaviour Science* 70 (2001): 275–84.

Foer, Jonathan Safran. *Eating Animals*. Little, Brown, 2009.

Forman, Richard T. T., and Lauren E. Alexander. "Roads and Their Major Ecological Effects." *Annual Review of Ecology and Systematics* 29 (1998): 201–33.

Foundation for Biomedical Research. "Animals Behind Top Drugs." Accessed March 18, 2025. https://fbresearch.org/medical-advances/top-drugs.

Foundation for Biomedical Research. "New Billboards Ask the Public to Decide Who They Would Rather See Live." April 5, 2011. https://www.prnewswire.com/news-releases/new-billboards-ask-the-public-to-decide-who-they-would-rather-see-live-a-rat-or-a-little-girl-119271574.html.

Fredrickson, T. N. "Ovarian Tumors of the Hen." *Environmental Health Perspectives* 73 (1987): 35–51.

Freund, Peter, and George Martin. "The Commodity That Is Eating the World: The Automobile, the Environment, and Capitalism." *Capitalism Nature Socialism* 7, no. 4 (1996): 3–29.

Gan, Elaine, Anna Tsing, Heather Swanson, and Nils Bubandt. "Haunted Landscapes of the Anthropocene." In *Arts of Living on a Damaged Planet*, edited by Anna Tsing, Heather Swanson, Elaine Gan, and Nils Bubandt, G1–G14. University of Minnesota Press, 2017.

Garner, Joseph P. "The Significance of Meaning: Why Do over 90% of Behavioral Neuroscience Results Fail to Translate to Humans, and What Can We Do to Fix It?" *ILAR Journal* 55, no. 3 (2014): 438–56.

German, Danielle, and Carl A. Latkin. "Exposure to Urban Rats as a Community Stressor Among Low-Income Urban Residents." *Journal of Community Psychology* 44, no. 2 (2016): 249–62.

Giles, James R., Robert G. Elkin, Lindsey S. Trevino, Mary E. Urick, Ramesh Ramachandran, and Patricia A. Johnson. "The Restricted Ovulator Chicken: A Unique Animal Model for Investigating the Etiology of Ovarian Cancer." *International Journal of Gynecological Cancer* 20, no. 5 (2010): 738–44.

Giraud, Eva, and Gregory Hollin. "Care, Laboratory Beagles and Affective Utopia." *Theory, Culture and Society* 33, no. 42 (2016): 27–49.

Glick, Megan. "Animal Instincts: Race, Criminality, and the Reversal of the 'Human.'" *American Quarterly* 65, no. 3 (2013): 639–59.

Gluck, John P. *Voracious Science and Vulnerable Animals: A Primate Scientist's Ethical Journey*. University of Chicago Press, 2016.

Goldberg-Hiller, Jonathan, and Noenoe K. Silva. "Sharks and Pigs: Animating Hawaiian Sovereignty Against the Anthropological Machine." *South Atlantic Quarterly* 110, no. 2 (2011): 429–46.

Gordon, Avery. *Ghostly Matters: Haunting and the Sociological Imagination.* University of Minnesota Press, 2008.

Government Accountability Office. "National Institutes of Health: Assessing Efforts to Improve Animal Research Could Lead to Greater Human Health Benefits." Accessed April 15, 2025. https://www.gao.gov/products/gao-25-107140.

Gumbs, Alexis Pauline. *Undrowned: Black Feminist Lessons from Marine Mammals.* AK Press, 2020.

Hatch, Anthony Ryan. *Silent Cells: The Secret Drugging of Captive America.* University of Minnesota Press, 2019.

Heath, Joan. "Animals in Research: Zebrafish." *The Conversation*, May 8, 2013. https://theconversation.com/animals-in-research-zebrafish-13804.

Herrmann, Kathrin. "Refinement on the Way Towards Replacement: Are We Doing What We Can?" In *Animal Experimentation: Working Towards a Paradigm Change*, edited by Kathrin Herrmann and Kimberley Jayne, 3–64. Brill, 2019.

Hirsch, Veronica. "Detailed Discussions of Legal Protections of the Domestic Chicken in the United States and Europe." Animal Legal and Historical Center, Michigan State University College of Law. Accessed August 19, 2024. https://www.animallaw.info/article/detailed-discussion-legal-protections-domestic-chicken-united-states-and-europe.

Horowitz, Alexandra. *Being a Dog: Following the Dog into the World of Smell.* Large print edition. Thorndike Press, 2016.

Horowitz, Alexandra. *On Looking: Eleven Walks with Expert Eyes.* Scribner, 2013.

Horowitz, Alexandra, and Marc Bekoff. "Naturalizing Anthropomorphism: Behavioral Prompts to Our Humanizing of Animals." *Anthrozoös* 20, no. 1 (2007): 23–35.

Hubbard, Phil, and Andrew Brooks. "Animals and Urban Gentrification: Displacement and Injustice in the Trans-Species City." *Progress in Human Geography* 45, no. 6 (2021): 1490–511.

Hubbard, Tasha. "Buffalo Genocide in Nineteenth-Century North America: 'Kill, Skin, and Sell.'" In *Colonial Genocide and Indigenous North America*, edited by Laban Hinton, Andrew Woolford, and Jeff Benvenuto, 292–305. Duke University Press, 2015.

Humane Society of the United States. "What to Do About Wild Rats." Accessed August 20, 2024. https://www.humanesociety.org/resources/what-do-about-wild-rats.

I-90 Wildlife Bridges Coalition. "A Historic Coalition." Accessed April 14, 2025. https://i90wildlifebridges.org/.

I-90 Wildlife Bridges Coalition. "Project Funding." Accessed August 18, 2024. https://i90wildlifebridges.org/project-funding/.

1-90 Wildlife Bridges Coalition. "Project Info." Accessed August 18, 2024. https://i90wildlifebridges.org/.

Interrante, Joseph. "The Road to Autopia: The Automobile and the Spatial Transformation of American Culture." In *The Automobile and American Culture*, edited by David C. Lewis and Laurence Goldstein, 502–17. University of Michigan Press, 1983.

Ioannidis, John P. A. "Why Most Published Research Findings Are False." *PLoS Medicine* 19, no. 8 (2005): 696–701.

Isenberg, Andrew. *The Destruction of the Bison: An Environmental History, 1750–1920*. Cambridge University Press, 2000.

Jackson, Zakiyyah Iman. "Animal: New Directions in the Theorization of Race and Posthumanism." *Feminist Studies* 39, no. 3 (2013): 669–85.

Jackson, Zakiyyah Iman. *Becoming Human: Matter and Meaning in an Antiblack World*. New York University Press, 2020.

Johnson, Karen. "The Standard of Perfection: Thoughts About the Laying Hen Model of Ovarian Cancer." *Cancer Prevention Research* 2, no. 2 (2009): 97–99.

Johnson, Patricia A., Christine S. Stephens, and James R. Giles. "The Domestic Chicken: Causes and Consequences of an Egg a Day." *Poultry Science* 94, no. 4 (2015): 816–20.

jones, pattrice. "Eros and the Mechanisms of Eco-Defense." In *Ecofeminism: Feminist Intersections with Other Animals and the Earth*, edited by Carol J. Adams and Lori Gruen, 91–108. Bloomsbury, 2014.

jones, pattrice. "Harbingers of (Silent) Spring: Archetypal Avians, Avian Archetypes, and the Truly Collective Unconscious." *Spring: A Journal of Archetype and Culture* 83 (2010): 183–210.

jones, pattrice. *The Oxen at the Intersection*. Lantern Books, 2014.

jones, pattrice. "Property, Profit, and (Re)Production: A Bird's-Eye View." In *Animal Oppression and Capitalism*, edited by David Nibert, 2:31–48. Bloomsbury, 2017.

jones, pattrice. "Provocations from the Field—Derangement and Resistance: Reflections from Under the Glare of an Angry Emu." *Animal Studies Journal* 8, no. 1 (2019): 1–20.

jones, pattrice. "Queer Eros in the Enchanted Forest: The Spirit of Stonewall as Sustainable Energy." *QED: A Journal in GLBTQ Worldmaking* 6, no. 2 (2019): 76–82.

jones, pattrice. "Stomping with the Elephants: Feminist Principles for Radical Solidarity." In *Igniting a Revolution: Voices in Defense of Mother Earth*, edited by Steven Best and Anthony Nocella II, 319–33. AK Press, 2006.

Justice, Daniel Heath. "A Relevant Resonance: Considering the Study of Indigenous National Literatures." In *Across Cultures / Across Borders: Canadian Aboriginal and Native American Literatures*, edited by Paul

DePasquale, Renate Eigenbrod, and Emma LaRocque, 61–76. Broadview Press, 2010.

Kalueff, Allan V., Adam Michael Stewart, Evan J. Kyzar, et al. "Time to Recognize Zebrafish 'Affective' Behavior." *Behaviour* 149, no. 10–12 (2012): 1019–36.

Karlsson, Fredrik. "Critical Anthropomorphism and Animal Ethics." *Journal of Agricultural and Environmental Ethics* 25, no. 5 (2011): 707–20.

Karuka, Manu. *Empire's Tracks: Indigenous Nations, Chinese Workers, and the Transcontinental Railroad.* University of California Press, 2019.

Kaukeinen, Dale E. "Myths About Rodents." *Pest Management Professional,* May 1, 2008.

Kim, Claire Jean. *Dangerous Crossings: Race, Species, and Nature in a Multicultural Age.* Cambridge University Press, 2015.

King, Angela, and Kim Shepard. "2 Newborn Orcas Spotted in Puget Sound in the Same Month." *KUOW NPR,* September 24, 2020.

King County Health Department. "Canine Leptospirosis in King County—2016." Accessed May 14, 2025. https://cdn.kingcounty.gov/-/media/king-county/depts/dph/documents/health-safety/disease-illness/diseases-from-animals/leptospirosis-in-dogs.pdf.

King County Public Health. "Diseases from Rodents, Pocket Pets, and Rabbits." Accessed August 20, 2024. https://kingcounty.gov/depts/health/communicable-diseases/zoonotic/facts-resources/diseases-by-animal/pocket-pets.aspx.

KING Staff. "3 More Southern Resident Orcas Declared Dead by Center for Whale Research." *KING5,* August 6, 2019.

Kisiel, Emma. *At Rest.* Accessed August 21, 2024. http://www.emmakisiel.com/work#/at-rest/.

Klein, Naomi. "Dancing the World into Being: A Conversation with Idle No More's Leanne Simpson." *YES! Magazine,* March 6, 2013.

Knight, Andrew. "Animals in Research: Do the Costs Outweigh the Benefits?" *The Conversation,* August 6, 2013.

Ko, Syl. "Addressing Racism Requires Addressing the Situation of Animals." In *Aphro-ism,* edited by Aph Ko and Syl Ko, 44–49. Lantern Books, 2017.

Kroll, Gary. "An Environmental History of Roadkill: Road Ecology and the Making of the Permeable Highway." *Environmental History* 20 (2015): 4–28.

Legendre, Juliette. "The Degrowth Movement Challenges the Conventional Wisdom on Economic Health." September 3, 2018. https://inequality.org/great-divide/degrowth-movement-economic-health/.

Lipstein, Emily. "You're Not Mapping Rats, You're Mapping Gentrification." *Deadspin.* Accessed August 20, 2024. https://deadspin.com/you-re-not-mapping-rats-you-re-mapping-gentrification-1835005060/.

Lopez, Barry, and Robin Eschner. *Apologia.* University of Georgia Press, 1998.

Mamers, Danielle Taschereau. "Human-Bison Relations as Sites of Settler Colonial Violence and Decolonial Resurgence." *Humanimalia* 10, no. 2 (2019): 10–41.

Mapes, Lynda. "Grieving Mother Orca Falling Behind Family as She Carries Dead Calf for a Seventh Day." *Seattle Times*, July 30, 2018.

Mapes, Lynda. "Southern Resident Orca Matriarch J17 Continues to Decline, New Photos Show." *Seattle Times*, May 17, 2019.

Marino, Lori, Richard C. Connor, R. Ewan Fordyce, et al. "Cetaceans Have Complex Brains for Complex Cognition." *PLoS Biology* 5, no. 5 (2007): e139.

Marino, Lori, Chet C. Sherwood, Bradley N. Delman, Cheuk Y. Tang, Thomas P. Naidich, and Patrick R. Hof. "Neuroanatomy of the Killer Whale (*Orcinus Orca*) from Magnetic Resonance Images." *Anatomical Record* 281A, no. 2 (2004): 1256–63.

Marshik, Joel, Lyle Renz, Jim Sipes, Dale Becker, and Dale Paulson. "Preserving a Spirit of Place: U.S. Highway 93 on the Flathead Indian Reservation." *UC Davis: Road Ecology Center.* 2001. https://escholarship.org/uc/item/51f1hodf.

Marx, Karl. *Capital.* Penguin Classics, 1990.

Marzluff, John M., and Tony Angell. *Gifts of the Crow: How Perception, Emotion, and Thought Allow Smart Birds to Behave Like Humans.* Atria Books, 2013.

Marzluff, John M., and Tony Angell. *In the Company of Crows and Ravens.* Yale University Press, 2005.

McComb, Karen, Cynthia Moss, Sarah M. Durant, Lucy Baker, and Soila Sayialel. "Matriarchs as Repositories of Social Knowledge in African Elephants." *Science* 292, no. 2296 (2001): 491–94.

McKinley, Elizabeth Ann, and Linda Tuhiwai Smith. *Handbook of Indigenous Education.* Springer, 2019.

McManus, Rich. "Ex-Director Zerhouni Surveys Value of NIH Research." *NIH Record* 65, no. 13 (2013): 1–12.

Merriam-Webster Dictionary. "Consume." Accessed August 19, 2024. https://www.merriam-webster.com/dictionary/consume.

Merriam-Webster Dictionary. "Forget." Accessed November 13, 2021. https://www.merriam-webster.com/dictionary/forget.

Mikics, Eva, Johanna Baranyi, and Jozsef Haller. "Rats Exposed to Traumatic Stress Bury Unfamiliar Objects—a Novel Measure of Hyper-Vigilance in PTSD Models?" *Physiology and Behavior* 94, no. 3 (2008): 341–48.

Monahan, Linda. "Mourning the Mundane: Memorializing Road-Killed Animals in North America." In *Mourning Animals: Rituals and Practices Surrounding Animal Death*, edited by Margo DeMello, 151–57. Michigan State University Press, 2015.

Moreau, Jean-Luc. "Simulating the Anhedonia Symptom of Depression in Animals." *Dialogues in Clinical Neuroscience* 4, no. 4 (2002): 351–60.

Moreton-Robinson, Aileen. *The White Possessive: Property, Power, and Indigenous Sovereignty.* University of Minnesota Press, 2015.

Mountain, Michael. "How the First Orcas Were Captured and Sold to Marine Parks." Whale Sanctuary Project. November 5, 2016. https://whalesanctuaryproject.org/first-orcas-captured-sold-marine-parks/.

Nadasdy, Paul. "The Gift in the Animal: The Ontology of Hunting and Human-Animal Sociality." *American Ethnologist* 34, no. 1 (2007): 25–43.

Narayanan, Yamini. "Animating Caste: Visceral Geographies of Pigs, Caste, and Violent Nationalisms in Chennai City." *Urban Geography* 44, no. 10 (2021): 2185–205.

Narayanan, Yamini. "Cow Protection as 'Casteised Speciesism': Sacralisation, Commercialisation and Politicisation." *South Asia* 41 (2018): 331–51.

Narayanan, Yamini. "An Ecofeminist Politics of Chicken Ovulation: A Socio-Capitalist Model of Ability as Farmed Animal Impairment." *Hypatia: A Journal of Feminist Philosophy* 39, no. 3 (2024): 1–21.

National Institutes of Health Office of Laboratory Animal Welfare. "The IACUC." Accessed August 19, 2024. https://olaw.nih.gov/resources/tutorial/iacuc.htm.

Netz, Reviel. *Barbed Wire: An Ecology of Modernity.* Wesleyan University Press, 2004.

Odell, Jenny. *How to Do Nothing: Resisting the Attention Economy.* Melville House, 2019.

Online Etymology Dictionary. "Abstracted." Accessed March 13, 2025. https://www.etymonline.com/search?q=abstracted.

Orca Nation. "The Social Intelligence of Orcas and Communication—Orca Series II." October 10, 2019. https://orcanation.org/the-social-intelligence-of-orcas.

Perathoner, Simon, Maria Lorena Cordero-Maldonado, and Alexander D. Crawford. "Potential of Zebrafish as a Model for Exploring the Role of the Amygdala in Emotional Memory and Motivational Behavior." *Journal of Neuroscience Research* 94, no. 6 (2016): 445–62.

Physicians Committee for Responsible Medicine. "Asthma Research for the 21st Century." *Good Science Digest.* Accessed April 14, 2025. https://www.pcrm.org/news/good-science-digest/asthma-research-21st-century.

Probyn-Rapsey, Fiona. "Anthropocentrism." In *Critical Terms for Animal Studies,* edited by Lori Gruen, 47–63. University of Chicago Press, 2018.

Roeder, Mikelle. "How to Sex Baby Chicks." Purina Mills. Accessed August 19, 2024. https://www.purinamills.com/chicken-feed/education/detail/how-to-sex-baby-chicks.

Rose, Deborah Bird. "Dialogue." In *Manifesto for Living in the Anthropocene,* edited by Katherine Gibson, Deborah Bird Rose, and Ruth Fincher, 127–31. Punctum Books, 2015.

Rose, Deborah Bird. "In the Shadow of So Much Death." In *Animal Death*, edited by Jay Johnston and Fiona Probyn-Rapsey, 1–20. Sydney University Press, 2013.

Rozenbaum, Mia. "Could AI Replace Animal Research?" Accessed April 16, 2025. https://www.understandinganimalresearch.org.uk /news/could-ai-replace-animal-research.

Scherer, Roberta W., Joerg J. Meerpohl, Nadine Pfeifer, Christine Schmucker, Guido Schwarzer, and Erik von Elm. "Full Publication of Results Initially Presented in Abstracts." *Cochrane Database of Systematic Reviews* 20, no. 11 (2018): 1–565.

Schuerman, Matthew L. *Newcomers: Gentrification and Its Discontents*. University of Chicago Press, 2019.

Shanahan, Ed. "New York City Rats: They're in the Park, on Your Block and Even at Your Table." *New York Times*, November 5, 2021.

Sharp, Lesley A. *Animal Ethos*. University of California Press, 2018.

Sharpe, Christina. *In the Wake: On Blackness and Being*. Duke University Press, 2016.

Shotwell, Alexis. *Against Purity: Living Ethically in Compromised Times*. University of Minnesota Press, 2016.

Simmons, James Raymond. *Fins, Feathers and Fur on the Turnpike*. Christopher Publishing House, 1938.

Simpson, Leanne Betasamosake. *As We Have Always Done: Indigenous Freedom Through Radical Resistance*. University of Minnesota Press, 2017.

Soron, Dennis. "Road Kill: Commodity Fetishism and Structural Violence." *TOPIAS* 18 (2008): 107–25.

Srinivasan, Krithika, and Rajesh Kasturirangan. "Political Ecology, Development, and Human Exceptionalism." *Geoforum* 75 (2016): 125–28.

Stanescu, Vasile. "Selling Eden: Environmentalism, Local Meat, and the Postcommodity Fetish." *American Behavioral Scientist* 63, no. 9 (2019): 1120–36.

Sullivan, Megan. "How to Heal an Emotionally Traumatized Pet." *PetMD*, February 6, 2018.

TallBear, Kim. "An Indigenous Reflection on Working Beyond the Human/Not Human." *GLQ: A Journal of Gay and Lesbian Studies* 21, no. 2 (2015): 230–35.

Taylor, Chloë. "The Precarious Lives of Animals: Butler, Coetzee, and Animal Ethics (Judith Butler and J. M. Coetzee)." *Philosophy Today* 52, no. 1 (2008): 60–72.

Taylor, Katy, and Laura Rego Alvarez. "An Estimate of the Number of Animals Used for Scientific Purposes Worldwide in 2015." *Alternatives to Laboratory Animals* 47, no. 5–6 (2020): 196–213.

Thrush, Coll. *Native Seattle: Histories from the Crossing-Over Place*. University of Washington Press, 2008.

Tuck, Eve, and C. Ree. "Glossary of Haunting." In *Handbook of Autoeth-nography*, edited by Stacey Holman Jones, Tony E. Adams, and Carolyn Ellis, 639–58. Left Coast, 2013.

Tuck, Eve, and K. Wayne Yang. "Decolonization Is Not a Metaphor." *Decolonization: Indigeneity, Education and Society* 1, no. 1 (2012): 1–40.

United States Department of Agriculture. "Annual Report Summary: 2023." Accessed April 16, 2025. https://www.aphis.usda.gov/sites/default/files/fy2023-research-animal-use-summary.pdf.

United States Department of Agriculture. "Milk Production." Last updated February 14, 2024. https://www.usda.gov/sites/default/files/documents/2024AOF-dairy-outlook.pdf.

United States Federal Highway Administration. "America's Highways, 1776–1976: A History of the Federal-Aid Program." 1977. https://archive.org/details/americashighways00unit/page/18.

United States Federal Highway Administration. "Our Nation's Highways 2011." Office of Highway Policy Information: Highway Finance Data Collection. Last updated November 7, 2014. https://www.fhwa.dot.gov/policyinformation/pubs/hf/pl11028/chapter1.cfm.

United States Forest Service. "Snake Migration LaRue–Pine Hills." 2006. https://www.fs.usda.gov/Internet/FSE_DOCUMENTS/stelprdb5106391.pdf.

University of Washington Office of Animal Welfare. "Institutional Compliance." Accessed August 14, 2025. https://sites.uw.edu/oawrss/office-of-animal-welfare/institutional-compliance/#regulatory-information-4.

Voss, Barbara L., Sue Fawn Chung, Kelly J. Dixon, et al. "The Archaeology of Precarious Lives: Chinese Railroad Workers in Nineteenth-Century North America." *Current Anthropology* 59, no. 3 (2018): 287–313.

Washington, Harriet A. *Medical Apartheid: The Dark History of Medical Experimentation on Black Americans from Colonial Times to the Present*. Doubleday, 2006.

Washington State Department of Health. "Leptospirosis." Accessed August 20, 2024. https://www.doh.wa.gov/youandyourfamily/illnessanddisease/leptospirosis.

Watson, L. A. "Remains to Be Seen." In *Economies of Death: Economic Logics of Killable Life and Grievable Death*, edited by Patricia J. Lopez and Kathryn Gillespie, 137–59. Routledge, 2015.

Watson, L. A. *The Roadside Memorial Project*. 2013. http://www.lawatsonart.com/the-roadside-memorial-project.html.

Wiesman, John, Kathy Lofy, Scott Lindquist, Jerrod Davis, Sherryl Terletter, and Marcia J. Goldoft. "Tularemia in Washington." *epiTrends* 22, no. 7 (2017): 1–4.

Willett, Cynthia. *Interspecies Ethics*. Columbia University Press, 2014.

Wolfe, Patrick. "Settler Colonialism and the Elimination of the Native." *Journal of Genocide Research* 8, no. 4 (2006): 387–409.

Index

and transportation, 42, 44–45, 63–64. *See also* capital; consumerism; consumption

captivity: bison in, 112; crows in, 4; lab animals in, 4, 68–70, 82–83; marine parks as sites of, 34; orcas in, 34; zoos as sites of, 34, 164–67

capture, 4, 34. *See also* captivity

care: for chickens, 115–17, 188; economic cost of, 117, 159; flourishing and, 16, 32, 94, 116–17, 145–46, 153, 156; harm and, 13, 15–16, 79–80; in lab research, 79–82; as mechanism of control, 79–80; narratives of, 120, 148; practicing acts of, 15, 28, 115, 178; veterinary, 117, 159

cars. *See* automobile

categories of life, 24–25, 29, 35, 52, 83, 129, 170, 180n6, 187n4

charismatic megafauna, 35, 84

chickens: at auctions, 105–7, 119; backyard, 96–101, 108–11, 114–16, 118; behavior, 100, 114–15; breeding of, 109–10; care for, 99–100, 115–18, 188; commodification of, 107; as food, 100–101, 107, 109, 115, 118; hormone implants for, 187n23; lack of legal protections for, 104, 184n3, 187n4; mourning, 116; ovarian cancer in, 108–9, 111; rats and, 130; reproduction, 107–10; as symbols of gentrification, 130; urban zoning for, 128–29. *See also* chicks; eggs; ovulation

chicks, 96–100, 107; grouse, 176; sexing of, 98–99; surplus of male, 109–10. *See also* chickens

chipmunk, 50–52, 174

cities. *See* Seattle; urban space

City Is More Than Human, The, 126, 130

civilization, 126, 128–30

climate crisis, 10, 30–32, 34

Coast Salish Tribes, 30, 126, 142

collisions, with animals, 53–54, 58–59; economic costs of, 60, 62; prevention of, 56–57, 59–67; safety and, 46, 48; as threat to property, 48

colonialism: definition of, 21–25; dismantling of, 29, 65–67; extraction and, 188n24; gentrification and, 125, 130, 140; hierarchy and, 23–24; inattentiveness and, 28; knowledge systems and, 144–45; property and, 142–43; settler, 23–24, 43–45, 65, 111–14, 125–26; transportation and, 42–45, 65; undoing, 65–67, 143–45

commodification: colonialism and, 24; consumption and, 102, 116–19; of death, 22, 24; of labor, 22, 24; of life, 22, 24–25, 97–98, 102–7; logic of, 21–22, 119–20; property and, 24–25, 113; reproduction and, 22, 102, 107–11, 163. *See also* capital; capitalism; commodities

commodities: animals as, 24, 98, 106–7, 116–19, 155; automobiles as, 45–47, 63–64; circulation of, 42, 45, 63–64, 106–7; definition of, 21–22, 46; inanimate, 46, 105; as property, 24; rendering, 22. *See also* capital; capitalism; commodification

commodity fetish, 46–47

competition, 64, 103, 119

complicity. *See* culpability

Confederated Salish and Kootenai Tribes, 65–66, 143

connection: allowing distance as, 178; desire for, 20, 73, 164–65; harm of seeking, 164–65; to others, 25, 32, 80, 107, 124, 164–65, 174; to place, 32, 127, 152; superficial, 164–65

connectivity: habitat, 60–61; transportation and, 44, 46

consent, 22, 71–72

conservation: of endangered species, 59, 62–63; harm in, 4; of species and individuals, 52, 63

consumerism: culture of, 28, 46–47, 63–64; hollowness of, 20–21; in human-animal relations, 20–22. *See also* capitalism; commodification; commodities; consumption

consumption: of animals, 21, 100–102, 107, 109–11, 116, 152, 188n24; automobiles and, 47, 63–64; definition of, 101–2, 114; futures not oriented around, 116, 118; inattentiveness and, 28; logic of, 21–22, 103, 118–20, 188n24; as possession, 114, 119–20 counting, 41, 72–73

cow-calf operations, 151–52, 163, 169. *See also* ranching

cows: branding of, 160; commodified for dairy, 109–11, 163; environmental harm and, 152, 156; as ghosts, 151; herding, 147–48; as occupiers, 113, 126; out of place, 155–56; as property, 153; raised for beef, 147–48, 151–52, 160, 163–64; reproduction in, 78, 109–10, 151, 163; in sanctuaries, 153–56; self-determination and, 162; social bonds in, 150–51, 168; solidarity with, 167; trauma in, 169. *See also* beef, free range; cow-calf operations; milk

Cow with Ear Tag #1389, The (Gillespie), 109

crows, 2–9, 13–15; behavior of, 2–4; and facial recognition, 3–4, 13; lab research on, 4; as pests, 6–7

culpability, 30, 41–42, 49, 58–59; acknowledgment of, 49, 55, 63, 140

curiosity, 11–14, 16, 95, 161–62, 173–74

dairy. *See* milk

damage: collateral, 45; ecosystem, 152, 156; property, 56, 62; undoing, 63, 65

dams, 157–58

death: accidental, 41–42; of animals on roads, 41–42, 50–56, 59, 62–63; commodification of, 22, 24; of farmed animals, 97, 111, 116, 148; of lab animals, 81, 185n18; mourning, 8, 33, 36, 55–56, 116; of rats, 136–37, 139

decolonization, 65–67, 143–44

degrowth, 64

dehumanization, 180n6

depression, in animals, 85

desire: for connection, 73, 164–65, 174; and consumption, 103, 115, 119–20, 137; human, 140, 178; to know, 15, 20, 174

development: medical, 71, 88, 91, 108, 185n23; suburban, 47, 62; transportation system, 42–46, 62–63; urban, 62, 123–25, 127–28, 164

digital sphere, 6–7

discomfort: with death, 52, 57; about eating animals, 116; with privilege, 64, 67; with violence, 41, 57, 140

disgust, 10, 40, 55

displacement: of animals, 26, 112–13, 127, 133, 139; definition of, 124; of humans, 142–43; multispecies, 23, 45, 121–24

disposability, 56, 62, 97–98, 119, 139

disposal, 41, 97, 139

dispossession: cities as sites of, 126; land, 22, 43–44, 141–42, 155, 188n24; possession and, 114; undoing, 141–42

dogs. *See* beagles

domestication, 110–11, 126, 128–30, 170

Donaldson, Sue, 167

Duwamish: river, 127–28; tribe, 30, 124, 128, 143

Eating Animals (Foer), 42

eggs: breeding of hens for, 109–10; as commodities, 105, 107; feeding as act of care, 115; harm in production of, 104, 188n24. *See also* chickens; chicks

Elwha Dam. *See* dams

emotion: animal, 35–37, 73, 80, 84–85, 107, 110, 169–70; burying, 160; paying attention and, 10; as response, 40, 42, 52, 63, 93. *See also* grief; shame

emus, 167–68

entitlement: to animal lives, 71, 101, 115–16; eating animals as act of, 115–16, 120; Manifest Destiny, 44; to space, 29, 49, 52, 125, 127, 130, 140–43; surrendering, 63, 67, 71, 93–94, 120, 116, 141, 146; to transportation, 52, 58

environmental destruction, 30, 64, 121–23, 127–28; and animal agriculture, 155–56; capitalism and, 46, 63–64, 188n24; from dams, 157–58; extreme weather and, 31; and marine life, 33–34; settler colonialism and, 126–27, 188n24; transportation and, 45–46, 49, 63, 65–66. *See also* climate crisis

eradication, 7, 16, 129, 135

pricing, of animals, 98, 105
primates, 72, 83–84, 87, 184n3, 185n18
Princess Angeline (orca), 34
privilege, 67, 83, 87; of forgetting, 139. *See also* entitlement
property, 23–24; animals as, 24, 102, 113–14, 141–42, 153; animals as owners of, 141–42; humans as, 24; markers of, 143; private, 24; undoing relationships of, 144. *See also* capital; ownership; possession, logic of
Puget Sound, 18, 22–23, 30, 33, 126

questioning, 12–13

race: hierarchies of, 24, 179–80n6; medical experimentation and, 71–72; restrictive covenants and, 128–29; superiority and, 22, 24
railroads, 42–47, 182n13
ranching, 43, 151–52, 158–60. *See also* animal agriculture
rats: behavior in, 84, 130–31; disease transmission and, 132–35; extermination of, 136–40; ignorance about, 130–35; in lab research, 73, 84–85, 136–37; in New York City, 134; as pests, 129–35
real estate: development, 121–23, 128; housing prices, 123
Real Rent, 143–44
reciprocity, 188n24
Ree, C., 27
remains, 39–41, 49, 51, 100–101, 111
remedy, 42. *See also* repair
remembering, 27, 42, 50, 140. *See also* forgetting; memory
rendering, 22, 40, 53, 111
repair, 15, 42, 63–67, 115, 144–46, 168; of ecosystems, 128, 157
reproduction, 22, 78, 102, 107–11, 116, 151, 166, 188n24. *See also* breeding
respect, 8, 15, 128, 161, 169, 178, 188n24
responsibility, 188n24; for harm, 29, 41–42, 59, 138, 140
rhinos, 166–67
roads: capitalism and, 42, 44–45, 63–64; colonialism and, 42–45, 63; construction of, 42–46; endangered species threatened by,

59, 62–63; entitlement to, 49, 52; expansion of, 48, 60–62; history of, 43–48; interstate, 48; mitigation projects for, 59–61, 63, 65–66; mundaneness of, 48, 50; safety of, 46, 48, 62; settler colonialism and, 42–44; signs on, 56–57; slavery and, 45; turnpike, 44; wildlife crossings on, 59–62, 65–66
Roadside Memorial Project, The (Watson), 56–57
road trips, 49–50
Rose, Deborah Bird, 18, 29
rupture, 34, 101–2, 118–19, 124, 127
rural culture, 158–59

sacrifice, 83, 87
salmon: Chinook, 33–34; spawning, 158
sanctuary, 117–18, 153–55
scent, 40, 69–70, 75, 94–95
school, 145
Schuerman, Matthew L., 142
Seattle, 1, 30–31, 121; gentrification in, 123, 126, 140, 142; history of, 126–29; rats in, 133–34, 139; real estate, 122, 140
Seattle Times, 35
SeaWorld, 34
self-determination, 67, 153, 162, 169, 178. *See also* agency; autonomy, lack of
sensory: experience of paying attention, 9; overload, 105–6; as transformative, 75–76, 94–95
settler colonialism, 23, 25, 43–45, 111–14, 125; animal agriculture and, 113, 187–88n24; fencing and, 113–14; gentrification, 125–26, 130, 140; and gold mining, 148–49; inattentiveness and, 28; permanence of, 113, 125–26; undoing of, 144–45. *See also* colonialism
settler subjects, 66–67, 143–45
shame, 41–42, 140
Shawnee National Forest, 59
Shotwell, Alexis, 121, 146
Simpson, Leanne Betasamosake, 30, 96, 144, 188n24
situatedness, 29, 158
slavery, 24, 45, 71, 182n13

slowing down, 8, 20; on roads, 52, 57, 62, 64
Smith, Linda Tuhiwai, 25
social conditioning, 131, 158–60
solace, 31–32
solastalgia, 31–33
solidarity, 168
Soron, Dennis, 47–49
Southern Residents (orcas), 33–36
sovereignty: animal, 66–67, 144; Native, 65–66, 143–44; settler, 125. *See also* self-determination
speed: of consumption, 103; of production, 46, 64; of travel, 41, 46, 48, 57, 60, 63–64
Spirit of Place, 65–66
stewardship, 23, 143
storytelling, 16
strangeness, 27
structures of power. *See* capitalism; colonialism; settler colonialism
subhuman, 24, 180n6
suffering: of animals killed on roads, 41; of bison, 45; desensitization to, 159–60; of dogs, 74; of farmed animals, 117–18, 160; of lab animals, 82–85, 92; of rats, 136–39; visibility of, 161

Tahlequah (orca), 33–38
TallBear, Kim, 23
tarantulas, 57–58
Taylor, Chloë, 85
technology: breeding as, 78–79; of fencing, 113–14; medical, 90, 93, 185n23
time: expectations of, 45, 103; shrinking, 45
Ti-Tahlequah (orca), 33
transformation: healing as, 75, 83, 95; of human-animal relations, 25, 63, 66–67, 94–95, 115–20, 141; land, 23, 45 46, 123, 126–27; of society, 48, 64–67
transportation systems, 42–48, 62
trauma: in bison, 45, 112; in dogs, 74–75; in elephants, 168–69; in farmed animals, 152, 168–69; healing from, 74–76, 83, 95, 117–18; intergenerational, 34, 45, 112, 168–69; in lab animals, 82–83; in orcas, 34

trust, 163, 170, 174
Tuck, Eve, 27, 125
Tuskegee Syphilis Study, 71–72

understanding: attempts at, 3–4, 8, 10–11, 37, 118, 150, 161–62, 174; intimacy in, 14–15; limits of, 3–4, 11, 15, 35–38, 161, 170
Undrowned (Gumbs), 145
United States Department of Agriculture (USDA): and animal research, 73; lab inspections, 81, 185n18
University of Washington: Center for Conservation Biology, 35; crow study, 4; medical research on animals, 68–71, 80–82, 88–89, 185n18, 185n20
urban space: animal agriculture in, 126, 128–29; development of, 121–23, 128; ecosystems and, 26, 121–22; greening of, 141; restrictive covenants in, 128–29; and settler histories, 126–30; zoning of, 128–29. *See also* gentrification; habitat; wildlife
use, animal: as food, 52–53, 101; for science, 82, 86–87, 93

value: cultural capital and, 106; differential, 24–25, 56, 179–80n6; economic, 22, 24–25, 44, 105, 111; of a life, 92, 98, 105; moral, 24–25
VINE Sanctuary, 153–56
violence: acknowledging, 26, 43, 56; car culture as, 41–42, 49, 59; culpability for, 41, 59, 140; desensitization to, 159–60; farming animals as, 137, 160, 162–63; invisible, 21, 139; mundane, 21, 18, 28, 42, 101–2; normalized, 21, 49, 81, 101–2, 137–38, 159–60; of property relations, 141–42; settler colonial, 23, 43, 65, 112, 126, 130, 142; stemming from ignorance, 134–35; structures of power and, 16, 21, 23, 29, 32, 92; undoing, 15, 65, 161
vulnerable populations, 72, 33, 72, 82–83, 91–92, 182n13

waste: animal remains as, 51–53, 110; consumption and, 101; eating to avoid, 51; of life, 52; of resources, 52

www.ingramcontent.com/pod-product-compliance
Lightning Source LLC
Jackson TN
JSHW022345081225
94419JS00001B/1